Juan D. Velásquez and Lakhmi C. Jain (Eds.)

Advanced Techniques in Web Intelligence – 1

T0180581

Studies in Computational Intelligence, Volume 311

Editor-in-Chief

Prof. Janusz Kacprzyk
Systems Research Institute
Polish Academy of Sciences
ul. Newelska 6
01-447 Warsaw
Poland
E-mail: kacprzyk@ibspan.waw.pl

Further volumes of this series can be found on our
homepage: springer.com

Vol. 288. I-Hsien Ting, Hui-Ju Wu, Tien-Hwa Ho (Eds.)
Mining and Analyzing Social Networks, 2010
ISBN 978-3-642-13421-0

Vol. 289. Anne Håkansson, Ronald Hartung, and
Ngoc Thanh Nguyen (Eds.)
*Agent and Multi-agent Technology for Internet and
Enterprise Systems,* 2010
ISBN 978-3-642-13525-5

Vol. 290. Weiliang Xu and John Bronlund
Mastication Robots, 2010
ISBN 978-3-540-93902-3

Vol. 291. Shimon Whiteson
Adaptive Representations for Reinforcement Learning, 2010
ISBN 978-3-642-13931-4

Vol. 292. Fabrice Guillet, Gilbert Ritschard,
Henri Briand, Djamel A. Zighed (Eds.)
Advances in Knowledge Discovery and Management, 2010
ISBN 978-3-642-00579-4

Vol. 293. Anthony Brabazon, Michael O'Neill, and
Dietmar Maringer (Eds.)
Natural Computing in Computational Finance, 2010
ISBN 978-3-642-13949-9

Vol. 294. Manuel F.M. Barros, Jorge M.C. Guilherme, and
Nuno C.G. Horta
*Analog Circuits and Systems Optimization based on
Evolutionary Computation Techniques,* 2010
ISBN 978-3-642-12345-0

Vol. 295. Roger Lee (Ed.)
*Software Engineering, Artificial Intelligence, Networking and
Parallel/Distributed Computing,* 2010
ISBN 978-3-642-13264-3

Vol. 296. Roger Lee (Ed.)
*Software Engineering Research, Management and
Applications,* 2010
ISBN 978-3-642-13272-8

Vol. 297. Tania Tronco (Ed.)
New Network Architectures, 2010
ISBN 978-3-642-13246-9

Vol. 298. Adam Wierzbicki
Trust and Fairness in Open, Distributed Systems, 2010
ISBN 978-3-642-13450-0

Vol. 299. Vassil Sgurev, Mincho Hadjiski, and
Janusz Kacprzyk (Eds.)
Intelligent Systems: From Theory to Practice, 2010
ISBN 978-3-642-13427-2

Vol. 300. Baoding Liu (Ed.)
Uncertainty Theory, 2010
ISBN 978-3-642-13958-1

Vol. 301. Giuliano Armano, Marco de Gemmis,
Giovanni Semeraro, and Eloisa Vargiu (Eds.)
Intelligent Information Access, 2010
ISBN 978-3-642-13999-4

Vol. 302. Bijaya Ketan Panigrahi, Ajith Abraham,
and Swagatam Das (Eds.)
Computational Intelligence in Power Engineering, 2010
ISBN 978-3-642-14012-9

Vol. 303. Joachim Diederich, Cengiz Gunay, and
James M. Hogan
Recruitment Learning, 2010
ISBN 978-3-642-14027-3

Vol. 304. Anthony Finn and Lakhmi C. Jain (Eds.)
Innovations in Defence Support Systems –1, 2010
ISBN 978-3-642-14083-9

Vol. 305. Stefania Montani and Lakhmi C. Jain (Eds.)
Successful Case-Based Reasoning Applications – 1, 2010
ISBN 978-3-642-14077-8

Vol. 306. Tru Hoang Cao
Conceptual Graphs and Fuzzy Logic, 2010
ISBN 978-3-642-14086-0

Vol. 307. Anupam Shukla, Ritu Tiwari, and Rahul Kala
Towards Hybrid and Adaptive Computing, 2010
ISBN 978-3-642-14343-4

Vol. 308. Roger Nkambou, Jacqueline Bourdeau, and
Riichiro Mizoguchi (Eds.)
Advances in Intelligent Tutoring Systems, 2010
ISBN 978-3-642-14362-5

Vol. 309. Isabelle Bichindaritz, Lakhmi C. Jain, Sachin Vaidya,
and Ashlesha Jain (Eds.)
Computational Intelligence in Healthcare 4, 2010
ISBN 978-3-642-14463-9

Vol. 310. Dipti Srinivasan and Lakhmi C. Jain (Eds.)
Innovations in Multi-Agent Systems and Applications – 1,
2010
ISBN 978-3-642-14434-9

Vol. 311. Juan D. Velásquez and Lakhmi C. Jain (Eds.)
Advanced Techniques in Web Intelligence – 1, 2010
ISBN 978-3-642-14460-8

Juan D. Velásquez and Lakhmi C. Jain (Eds.)

Advanced Techniques in Web Intelligence – 1

 Springer

Dr. Juan D. Velásquez
Department of Industrial Engineering
School of Engineering and Science
University of Chile Republica 701
Santiago, P.C. 837-0720
Chile
E-mail: jvelasqu@dii.uchile.cl

Prof. Lakhmi C. Jain
School of Electrical and Information Engineering
University of South Australia
Adelaide
Mawson Lakes Campus
South Australia
Australia
E-mail: Lakhmi.jain@unisa.edu.au

ISBN 978-3-642-26498-6   ISBN 978-3-642-14461-5 (eBook)

DOI 10.1007/978-3-642-14461-5

Studies in Computational Intelligence   ISSN 1860-949X

Typeset & Cover Design: Scientific Publishing Services Pvt. Ltd., Chennai, India.

Printed on acid-free paper

9 8 7 6 5 4 3 2 1

springer.com

Preface

The term Web Intelligence is defined as a new line of scientific research and development, which is used to explore the fundamental roles and practical impact of Artificial Intelligence together with advanced Information Technology and its effect on the future generations of Web-empowered products. These include systems, services, amongst other activities, all of which are carried out by the Web Intelligence Consortium (http://wi-consortium.org/).

Web Intelligence was first coined in the late 1999's. From that time, many new algorithms, methods and techniques were developed and used extracting both knowledge and wisdom from the data originating from the Web. A number of initiatives have been adopted by the world communities in this area of study. These include books, conference series, and journals. This latest book encomposes a variaty of up to date state of the art approaches in Web Intelligence. Furthermore, it hightlights successful applications in this area of research within a practical context.

The present book aims to introduce a selection of research applications in the area of Web Intelligence. We have selected a number of researchers around the world, all of which are experts in their respective research areas. Each chapter focuses on a specific topic in the field of Web Intelligence. Furthermore the book consists of a number of innovative proposals which will contribute to the development of web science and technology for the long-term future, rendering this collective work a valuable piece of knowledge. It was a great honour to have collaborated with this team of very talented experts. We also wish to express our grattitude to those who reviewed this book offering their constructive feedbacks.

Our thanks are also due to the Springer-Verlag staff for their excellent support given to us during the preparation of this manuscript.

We are also indebted to the staff of the Millennium Institute on Complex Engineering Systems (ICM: P-05-004-F, CONICYT: FBO16), which also partially funded this work.

Santiago, Chile,   Juan D. Velásquez
Adelaide, Australia   Lakhmi C. Jain
June 2010

Contents

List of Contributors

Felipe Aguilera
Department of Compter Science,
University of Chile,
Av. Blanco Encalada 2120,
Santiago, Chile
faguiler@dcc.uchile.cl

Ricardo Baeza-Yates
Yahoo! Research, Barcelona, Spain,
rbaeza@acm.org

Paolo Boldi
Dip. di Scienze dell'Informazione,
Univ. degli Studi di Milano, Italy
boldi@dsi.unimi.it

Robert F. Dell
Naval Postgraduate School,
Operations Research Department,
Monterey, California, USA
dell@nps.edu

José Manuel Gómez-Pérez
iSOCO, Intelligent Software
Components S.A
jmgomez@isoco.com

Zdeněk Horák
Faculty of Electrical Engineering
and Computer Science,

VŠB–Technical
University of Ostrava,
Czech Republic
zdenek.horak.st4@vsb.cz

Lakhmi C. Jain
KES Centre,
School of Electrical and
Information Engineering,
University of South Australia,
Adelaide, Mawson Lakes Campus,
South Australia SA 5095, Australia
Lakhmi.Jain@unisa.edu.au

Angel Jimenez-Molina
KAIST, Korea Advanced Institute of
Science and Technology,
335 Gwahangno
(373-1 Guseong-dong),
Yuseong-gu, Daejeon (305-701),
Republic of Korea
anjimenez@kaist.ac.kr

Jason J. Jung
Knowledge Engineering Laboratory,
Yeungnam University, Gyeongsan,
Republic of Korea
j2jung@intelligent.pe.kr,
ynu.ac.kr. gmail.com

Gastón L'Huillier
Web Intelligence Research Group,
Department of Industrial
Engineering
School of Engineering and Science,
University of Chile, República 701,
Santiago, Chile
glhuilli@dii.uchile.cl

In-Young Ko
KAIST, Korea Advanced Institute of
Science and Technology,
335 Gwahangno
(373-1 Guseong-dong),
Yuseong-gu, Daejeon (305-701),
Republic of Korea
iko@kaist.ac.kr

Hyung-Min Koo
KAIST, Korea Advanced Institute of
Science and Technology,
335 Gwahangno
(373-1 Guseong-dong),
Yuseong-gu, Daejeon (305-701),
Republic of Korea
hmkoo@kaist.ac.kr

Miloš Kudělka
Faculty of Electrical Engineering and
Computer Science, VŠB–Technical
University of Ostrava,
Czech Republic
kudelka@vsb.cz

Víctor L. Rebolledo
Web Intelligence Research Group,
Department of Industrial
Engineering
School of Engineering and Science,
University of Chile,
República 701,
Santiago, Chile
vireboll@ing.uchile.cl

Pablo E. Román
Web Intelligence Research Group,
Department of Industrial
Engineering
School of Engineering and Science,
University of Chile,
República 701,
Santiago, Chile
proman@ing.uchile.cl

Sebastián A. Ríos
Web Intelligence Research Group,
Department of Industrial
Engineering
School of Engineering and Science,
University of Chile,
República 701,
Santiago, Chile
srios@dii.uchile.cl

Carlos Ruiz
iSOCO, Intelligent Software
Components S.A
cruiz@isoco.com

Václav Snášel
Faculty of Electrical Engineering and
Computer Science,
VŠB–Technical
University of Ostrava,
Czech Republic
vaclav.snasel@vsb.cz

Juan D. Velásquez
Web Intelligence Research Group,
Department of
Industrial Engineering
School of Engineering and Science,
University of Chile,
República 701, Santiago, Chile
jvelasqu@dii.uchile.cl

Chapter 1
Innovations in Web Intelligence

Gastón L'Huillier, Juan D. Velásquez, and Lakhmi C. Jain

Abstract. The information footprints of a rapidly increasing influx of Internet users present us with an immense source of information that ultimately contributes to the construction of innovative web technology suitable for the future generations. Likewise, Web Intelligence has been presented as the usage of advanced techniques in Artificial Intelligence and Information Technology for the purpose of exploring, analysing, and extracting knowledge from Web data. In this chapter, the use of Web Intelligence is discussed together with ways in which a wide range of research is benefiting this area for the long-term. Also the books' purpose and structure are introduced, together with a list of resources to explore this area of research further.

1.1 Introduction

Web Intelligence has been considered during the last decade as one of the leading areas of research and development in modern science. Ever since the Web was invented by Tim Berners-Lee [3], data about human behaviour and activities has been gathered at different levels. This is specially in terms of their interests when they are arranged to follow a link, the buyers of a specific product, or the way in which they feel about a specific topic in a virtual community. This behaviour has left a footprint that must be considered for further analysis. This information, keeps feeding the Web constantly and which enable us to explore the the dynamics of our society, future trends in various aspects of our every-days life, and other questions which are as yet beyond our imagination.

Gastón L'Huillier · Juan D. Velásquez
Web Intelligence Research Group, University of Chile,
Department of Industrial Engineering, Repblica 701, Santiago, Chile
e-mail: jvelasqu@dii.uchile.cl, glhuilli@dii.uchile.cl

Lakhmi C. Jain
KES Centre, School of Electrical and Information Engineering, University of South Australia, Adelaide, Mawson Lakes Campus, South Australia SA 5095, Australia
e-mail: lakhmi@unisa.edu.au

J.D. Velásquez and L.C. Jain (Eds.): Advanced Techniques in Web Intelligence – 1, SCI 311, pp. 1–17.
springerlink.com   © Springer-Verlag Berlin Heidelberg 2010

The rapid growth of the World Wide Web, the assembly of large scale volumes of web data, and ever exponentially increasing applications has lead to the development of ever smarter approaches to extract patterns and build knowledge with the aide of artificial intelligence techniques. These techniques have been used, together with information technology, in a wide range of applications. This is where semantics, social network analysis, web structure, content, usage, and other aspects have already been and will increasingly keep being included in many application domains.

To keep up-to-date in the research areas of Web Intelligence is fundamental to further contribute towards the understanding of how the Web can improve our everyday life. This is the goal of this book, which is to present advanced techniques in Web Intelligence, show their main contributions, applications, and limitations. This book can be considered as a compendium of today's techniques that are likely to continue in the development of independent research of areas. Together these represent what the Web Intelligence concept stand for that is; to explore the fundamental roles and impacts of Artificial Intelligence and Information Technology for the next generations of Web-empowered products, Systems, Services, and Activities[1].

This chapter is structured as follows: First, in section 1.2 a brief overview of advanced techniques in Web Intelligence is presented, and different branches are discussed. Second, in section 1.3, all chapters included in this book are introduced, together with a discussion of their main characteristics. The summary of chapter is given in Section 1.4. A list of resources is provided in Section 1.5.

1.2   An Overview of the Advanced Techniques Used in Web Intelligence

Web Intelligence covers a wide area here artificial intelligence and information technology are integrated to enhance different web-based applications. Different techniques and technologies have been used by researchers and practitioners over the years. Concepts such as Web information repositories [25], Web user behaviour analysis [20, 23], Web content [15, 21] and structure mining [16], social network analysis [4], the semantic Web [17, 22]. In addition more general concepts such as Knowledge Discovery from Databases [7] and Knowledge Representation [5] are the key to understand the basics from which Web Intelligence has been assembled.

In terms of knowledge representation and storage, fields such as logic, ontology, and computation are critical in order to support the basic structure evolving from a Web of data to a Web of knowledge [24]. Furthermore, once knowledge is mined from the web data, different standards, such as the Predictive Model Mark-up Language (PMML) [18], have been developed to store and manage the different patterns extracted from the content. These repositories have been developed for use in Multidimensional Analysis architectures. This is where Extraction, Transformation, and Loading from web-based resources, Data Web-house Meta-data Modelling, OLAP queries, and its visualization have been extensively studied [19].

---

[1] As described by the WI consortium http://wi-consortium.org

As part of the collection, pre-processing, and cleaning of data, several issues on privacy and quality measures must be considered [24]. Different web mining applications, such as Web User Behaviour, Content of Different Web Sites, and the analysis of the web as a graph have been discussed in the areas of Web Intelligence, Data Mining, Machine Learning, Information Retrieval, and Artificial Intelligence communities in various conferences and journals (see section 1.5).

Applications oriented to the analysis of information preferences, web usability and usefulness considerations such as helping the web user to find information have been areas of intrust. They have found the centre of attention for web usage mining researchers [24]. Other applications, such as the identification of where, how, and items which must be considered in a particular content of a given web site has formed the focus for Web Content Mining researchers [6]. The structure, representation, and its analysis has been considered as part of Web structure mining [16] and the information retrieval [2]. In previous applications, traditional supervised and un-supervised machine learning algorithms [10, 14], and data quality, visualization, characterization, analysis techniques have been developed for the Web Intelligence Community [24].

In all of the later applications, the original Web data is presented in appropriate formats that must be processed and represented in terms for the technique to be used. In this context, Web logs, the Web-site contents, and the Hyperlink Structure of the Web, have been considered as the main source of information. Privacy issues on the sessionization process, such as using invasive tools to identify the users [24], and social network analysis where the user's contacts are exposed, have been the focus of further developments in privacy preserving data mining for Web Intelligence applications [1, 26].

One of the most promising research and application areas in Web Intelligence are the social networks and in web communities' analysis [8, 12, 17]. First studies on web structure has led to different ranking algorithms and techniques that are currently used in the analysis on how communities are formed. This includes the HITS algorithm, where authorities and hubs are identified [13]. Nowadays the content is not exclusively reserved for expert web-masters. The content on the Web is being developed by almost all of its users in web blogging, web forums, microblogging, virtual encyclopedias, social network applications. This enables the storage and generation of linked and structured information, that can be associated with text messages and multimedia information such as pictures and videos. All of these are currently being considered as a rich source of many research projects, where techniques such as social network analysis, text mining, and web mining are used together.

Finally, advances in Web Intelligence research are being focused on the enhancement of the semantic Web. The main objective is to provide a Web of descriptive meaning. There are different key aspects of knowledge representation such as computational linguistics, and other related Computer Science areas which have contributed to its development [22, 27]. Several standards for meta-data processing such as the Resource Description Framework (RDF) [11], Web Ontology Language (OWL) [9], and social network representations of RDF, such as Friend of a Friend

(FOAF) [8], have been proposed as contributions to semantics considerations in the Web.

1.3   Chapters Included in the Book

This book contains ten chapters and is edited using the contributions of various researchers and experts in the Web Intelligence field. In a broad perspective, this book includes topics such as Knowledge Representation and Pattern Extraction Storage, Web Content Mining for Information Granules (introduced as MicroGenres), Web Structure Mining, Web Usage Mining, Web Services Applications for Ubiquitous Computing, Ubiquitous Services in Social Networks, Ontology Engineering, and Web Intelligence in the Social Web.

Chapter two, Advanced Techniques in Web Data Pre-Processing and Cleaning by Pablo R. Roman, Robert F. Dell, and Juan D. Velasquéz, presents different approaches and issues regarding the pre-processing and cleaning of Web data. Different characteristics for different Web Intelligence, such as Web Structure Mining, Web Content Mining, and Web Usage Mining Applicatiions are discussed.

Chapter three, Web Pattern Extraction and Storage by Victor L. Rebolledo, Gastón L'Huillier, and Juan D. Velásquez, addresses juvenal different technology based architectures used for knowledge representation and pattern storage. Here, a large number of techniques for pattern extraction, such as Feature Selection and Extraction, Data Mining models, Model Assessment, and Performance Measures, from Web Data and its Multidimensional Storage by using PMML is presented.

Chapter four, Web Content Mining Using MicroGenres by Václav Snášel, Miloš Kudělka, and Zdeněk Horák, introduces an specific application of web content mining using MicroGenres, where specific components of a web page are identified and analysed.

Chapter five, Web Structure Mining by Ricardo Baeza-Yates and Paolo Boldi, presents basic properties, concepts, and models of the Web graph. Also, Developments in Link Ranking and Web Page Clustering are discussed, as well as Algorithmic issues as Streaming Computation on Graphs and Web graph Compression.

Chapter six, Web Usage Mining by Pablo E. Roman, Gastón L'Huillier, and Juan D. Velásquez, presents different techniques and issues regarding the characterization of the web user browser behaviour, as well as the representation of its preferences, and further techniques used for its Pattern Extraction. Finally, recent applications on Adaptive Web Sites, Web Personalization, and Recommendation are discussed.

Chapter seven, User-Centric Web Services for Ubiquitous Computing by In-Young Ko, Hyung-Min Koo, and Angel Jimenez-Molina, presents a novel application of Web Services in Ubitiqitous Computing in which essential requirements, current research on different frameworks, and a Task-Oriented Services Framework are discussed together with a demo application example.

Chapter eight, Ontological Engineering and the Semantic Web by José Manuel Gómez-Perez and Carlos Ruiz, discusses fundamental concepts on Knowledge Representation and Ontology Engineering, as well as a Methodological Approach to

Ontology Engineering, introduced as Methontology. Afterwards, a discussion on Reasoning, Modularization and Customization, Networked Ontologies, and Ontology development frameworks is overviewed, Applications such as Semantic web services, semantic applications in Public Administrations, semantic applications in eBusiness, and new challenges in the semantic cloud.

Chapter nine, Web Intelligence on the Social Web by Sebastián A. Ríos and Felipe Aguilera, presents an overview on how virtual communities and social networks could be analysed and how knowledge could be extracted. Also, different web mining techniques and how they could be applied to social network analysis introduced. A brief introduction on how web mining could be applied in Semantic Web Sites from a Social Network Analysis point of view is discussed.

The Final chapter, Intelligent Ubiquitous Services Based on Social Networks is authored by Jason J. Jung. He has presented two case studies related to the social network-based mobile recommendation services.

1.4   Summary

In this chapter broad areas of Web Intelligence have been discussed and analysed from this books perspective. A general overview of the book's chapters was introduced and a comprehensive list of the resources is presented. The remaining chapters will consider further details on recent advances in the area of Web Intelligence.

1.5   Resources

A sample of the resources is presented to explore the field of Web intelligence and intelligent systems further. First, a list of the main Journals in the field is given. Secondly, a list of the conferences and their proceedings are listed. Finally, a list related books is presented.

1.5.1   Journals

- IEEE Internet Computing, IEEE Computer Society Press, USA,
  www.computer.org/internet/
- AI Magazine, USA,
  www.aaai.org
- Web Intelligence and Agent Systems, IOS Press, The Netherlands,
  http://wi-consortium.org/journal.html
- International Journal of Knowledge and Web Intelligence (IJKWI), Inter-Science.

- International Journal of Knowledge-Based Intelligent Engineering Systems, IOS Press, The Netherlands,
  `www.kesinternational.org/journal/`
- IEEE Transactions on Knowledge and Data Engineering (TKDE), IEEE Computer Society Press, USA,
  `www.computer.org/tkde`
- Data & Knowledge Engineering (DKE), Elsevier Science Publishers B. V., The Netherlands,
  `www.elsevier.com/locate/datak`
- Knowledge-Based Systems, Elsevier Science Publishers B. V., The Netherlands,
  `www.elsevier.com/locate/knosys`
- Artificial Intelligence, Elsevier Science Publishers B. V., The Netherlands,
  `www.elsevier.com/locate/artint`
- Computer, IEEE Computer Society Press, USA,
  `www.computer.org/compute`
- Journal of Web Semantics, Elsevier Science Publishers B. V., The Netherlands,
  `http://www.elsevier.com/locate/websem`
- International Journal of Semantic Web and Information Systems, IGI Global
  `www.ijswis.org/`
- ACM Transactions on Internet Technology, ACM Press, USA,
  `http://toit.acm.org/`
- Communications of the ACM, ACM Press, USA,
  `http://cacm.acm.org/`
- IEEE Pervasive Computing, IEEE Computer Society Press, USA,
  `www.computer.org/pervasive/`
- IEEE Transactions on Systems, Man, and Cybernetics, IEEE Computer Society Press, USA,
  `http://www.ieeesmc.org/publications/index.html`
- ACM Computing Surveys, ACM Press, USA,
  `http://surveys.acm.org/`
- Knowledge and Information Systems, Springer Science+Business Media, USA,
  `www.cs.uvm.edu/~kais/`
- Data Mining and Knowledge Discovery, Springer Science+Business Media, USA
- Internet Mathematics, A K Peters ltd. Publishers of Science and Technology
  `www.internetmathematics.org/`
- Machine Learning, Springer Science+Business Media, USA
- Journal of Machine Learning Research, MIT Press, USA
  `http://jmlr.csail.mit.edu/`
- SIGKDD Explorations, ACM Press, USA,
  `www.sigkdd.org/explorations/`

1.5.2   Conferences

- IEEE/WIC/ACM International Conferences on Web Intelligence (WI)
- KES International Conference Series (KES)
- Australian World Wide Web Conferences
- ACM International Conferences on Web Search and Web Data Mining (WSDM)
- ACM Conferences on Information and Knowledge Management (CIKM)
- ACM International Conferences on World Wide Web (WWW)
- International Conferences on Very Large Data Bases (VLDP)
- International Conferences on Web Information Systems Engineering (WISE)
- ACM SIGKDD International Conference on Knowledge Discovery and Data Mining (KDD)
- ACM International Conferences on Machine Learning (ICML)
- IEEE International Conferences on Data Mining (ICDM)
- International ACM SIGIR Conferences on Research and Development in Information Retrieval (SIGIR)
- SIAM International Conference on Data Mining (SDM)
- Pacific-Asia Conferences in Advances in Knowledge Discovery and Data Mining (PAKDD)
- International Semantic Web Conferences (ISWC)
- International Joint Conference on Artificial Intelligence (IJCAI)

1.5.3   Conferences Proceedings

- Juan D. Velásquez, Sebastían A. Ríos, Robert J. Howlett, Lakhmi C. Jain (Eds.): Knowledge-Based and Intelligent Information and Engineering Systems, 13th International Conference, KES 2009, Santiago, Chile, September 28-30, 2009, Proceedings, Part I. Lecture Notes in Computer Science 5711 Springer 2009
- Juan D. Velásquez, Sebastían A. Ríos, Robert J. Howlett, Lakhmi C. Jain (Eds.): Knowledge-Based and Intelligent Information and Engineering Systems, 13th International Conference, KES 2009, Santiago, Chile, September 28-30, 2009, Proceedings, Part II. Lecture Notes in Computer Science 5712 Springer 2009
- Ignac Lovrek, Robert J. Howlett, Lakhmi C. Jain (Eds.): Knowledge-Based Intelligent Information and Engineering Systems, 12th International Conference, KES 2008, Zagreb, Croatia, September 3-5, 2008, Proceedings, Part I. Lecture Notes in Computer Science 5177 Springer 2008
- Ignac Lovrek, Robert J. Howlett, Lakhmi C. Jain (Eds.): Knowledge-Based Intelligent Information and Engineering Systems, 12th International Conference, KES 2008, Zagreb, Croatia, September 3-5, 2008, Proceedings, Part II. Lecture Notes in Computer Science 5178 Springer 2008
- Ignac Lovrek, Robert J. Howlett, Lakhmi C. Jain (Eds.): Knowledge-Based Intelligent Information and Engineering Systems, 12th International Conference, KES 2008, Zagreb, Croatia, September 3-5, 2008, Proceedings, Part III. Lecture Notes in Computer Science 5179 Springer 2008

- Bruno Apolloni, Robert J. Howlett, Lakhmi C. Jain (Eds.): Knowledge-Based Intelligent Information and Engineering Systems, 11th International Conference, KES 2007, XVII Italian Workshop on Neural Networks, Vietri sul Mare, Italy, September 12-14, 2007. Proceedings, Part I. Lecture Notes in Computer Science 4692 Springer 2007
- Bruno Apolloni, Robert J. Howlett, Lakhmi C. Jain (Eds.): Knowledge-Based Intelligent Information and Engineering Systems, 11th International Conference, KES 2007, XVII Italian Workshop on Neural Networks, Vietri sul Mare, Italy, September 12-14, 2007. Proceedings, Part II. Lecture Notes in Computer Science 4693 Springer 2007
- Bruno Apolloni, Robert J. Howlett, Lakhmi C. Jain (Eds.): Knowledge-Based Intelligent Information and Engineering Systems, 11th International Conference, KES 2007, XVII Italian Workshop on Neural Networks, Vietri sul Mare, Italy, September 12-14, 2007, Proceedings, Part III. Lecture Notes in Computer Science 4694 Springer 2007
- Bogdan Gabrys, Robert J. Howlett, Lakhmi C. Jain (Eds.): Knowledge-Based Intelligent Information and Engineering Systems, 10th International Conference, KES 2006, Bournemouth, UK, October 9-11, 2006, Proceedings, Part I. Lecture Notes in Computer Science 4251 Springer 2006
- Bogdan Gabrys, Robert J. Howlett, Lakhmi C. Jain (Eds.): Knowledge-Based Intelligent Information and Engineering Systems, 10th International Conference, KES 2006, Bournemouth, UK, October 9-11, 2006, Proceedings, Part II. Lecture Notes in Computer Science 4252 Springer 2006
- Bogdan Gabrys, Robert J. Howlett, Lakhmi C. Jain (Eds.): Knowledge-Based Intelligent Information and Engineering Systems, 10th International Conference, KES 2006, Bournemouth, UK, October 9-11, 2006, Proceedings, Part III. Lecture Notes in Computer Science 4253 Springer 2006
- Rajiv Khosla, Robert J. Howlett, Lakhmi C. Jain (Eds.): Knowledge-Based Intelligent Information and Engineering Systems, 9th International Conference, KES 2005, Melbourne, Australia, September 14-16, 2005, Proceedings, Part I. Lecture Notes in Computer Science 3681 Springer 2005
- Rajiv Khosla, Robert J. Howlett, Lakhmi C. Jain (Eds.): Knowledge-Based Intelligent Information and Engineering Systems, 9th International Conference, KES 2005, Melbourne, Australia, September 14-16, 2005, Proceedings, Part II. Lecture Notes in Computer Science 3682 Springer 2005
- Rajiv Khosla, Robert J. Howlett, Lakhmi C. Jain (Eds.): Knowledge-Based Intelligent Information and Engineering Systems, 9th International Conference, KES 2005, Melbourne, Australia, September 14-16, 2005, Proceedings, Part III. Lecture Notes in Computer Science 3683 Springer 2005
- Rajiv Khosla, Robert J. Howlett, Lakhmi C. Jain (Eds.): Knowledge-Based Intelligent Information and Engineering Systems, 9th International Conference, KES 2005, Melbourne, Australia, September 14-16, 2005, Proceedings, Part IV. Lecture Notes in Computer Science 3684 Springer 2005

- Mircea Gh. Negoita, Robert J. Howlett, Lakhmi C. Jain (Eds.): Knowledge-Based Intelligent Information and Engineering Systems, 8th International Conference, KES 2004, Wellington, New Zealand, September 20-25, 2004. Proceedings. Part I. Lecture Notes in Computer Science 3213 Springer 2004 '
- Mircea Gh. Negoita, Robert J. Howlett, Lakhmi C. Jain (Eds.): Knowledge-Based Intelligent Information and Engineering Systems, 8th International Conference, KES 2004, Wellington, New Zealand, September 20-25, 2004. Proceedings. Part II. Lecture Notes in Computer Science 3214 Springer 2004
- Mircea Gh. Negoita, Robert J. Howlett, Lakhmi C. Jain (Eds.): Knowledge-Based Intelligent Information and Engineering Systems, 8th International Conference, KES 2004, Wellington, New Zealand, September 20-25, 2004. Proceedings. Part III. Lecture Notes in Computer Science 3215 Springer 2004
- Vasile Palade, Robert J. Howlett, Lakhmi C. Jain (Eds.): Knowledge-Based Intelligent Information and Engineering Systems, 7th International Conference, KES 2003, Oxford, UK, September 3-5, 2003, Proceedings, Part I. Lecture Notes in Computer Science 2773 Springer 2003
- Vasile Palade, Robert J. Howlett, Lakhmi C. Jain (Eds.): Knowledge-Based Intelligent Information and Engineering Systems, 7th International Conference, KES 2003, Oxford, UK, September 3-5, 2003, Proceedings, Part II. Lecture Notes in Computer Science 2774 Springer 2003
- Proceedings of the 8th IEEE International Conference on Data Mining (ICDM 2008), December 15-19, 2008, Pisa, Italy. IEEE Computer Society 2008
- Proceedings of the 7th IEEE International Conference on Data Mining (ICDM 2007), October 28-31, 2007, Omaha, Nebraska, USA. IEEE Computer Society 2007
- David Wai-Lok Cheung, Il-Yeol Song, Wesley W. Chu, Xiaohua Hu, Jimmy J. Lin (Eds.): Proceedings of the 18th ACM Conference on Information and Knowledge Management, CIKM 2009, Hong Kong, China, November 2-6, 2009. ACM 2009
- James G. Shanahan, Sihem Amer-Yahia, Ioana Manolescu, Yi Zhang, David A. Evans, Aleksander Kolcz, Key-Sun Choi, Abdur Chowdhury (Eds.): Proceedings of the 17th ACM Conference on Information and Knowledge Management, CIKM 2008, Napa Valley, California, USA, October 26-30, 2008. ACM 2008
- Mário J. Silva, Alberto H. F. Laender, Ricardo A. Baeza-Yates, Deborah L. McGuinness, Bjrn Olstad, ystein Haug Olsen, Andr O. Falco (Eds.): Proceedings of the Sixteenth ACM Conference on Information and Knowledge Management, CIKM 2007, Lisbon, Portugal, November 6-10, 2007. ACM 2007
- Otthein Herzog, Hans-Jrg Schek, Norbert Fuhr, Abdur Chowdhury, Wilfried Teiken (Eds.): Proceedings of the 2005 ACM CIKM International Conference on Information and Knowledge Management, Bremen, Germany, October 31 - November 5, 2005. ACM 2005
- Proceedings of the 2001 ACM CIKM International Conference on Information and Knowledge Management, Atlanta, Georgia, USA, November 5-10, 2001. ACM 2001

- Ricardo A. Baeza-Yates, Paolo Boldi, Berthier A. Ribeiro-Neto, Berkant Barla Cambazoglu (Eds.): Proceedings of the Second International Conference on Web Search and Web Data Mining, WSDM 2009, Barcelona, Spain, February 9-11, 2009. ACM 2009
- Juan Quemada, Gonzalo Len, Yolle S. Maarek, Wolfgang Nejdl (Eds.): Proceedings of the 18th International Conference on World Wide Web, WWW 2009, Madrid, Spain, April 20-24, 2009. ACM 2009
- Klemens Böhm, Christian S. Jensen, Laura M. Haas, Martin L. Kersten, Per-ke Larson, Beng Chin Ooi (Eds.): Proceedings of the 31st International Conference on Very Large Data Bases, Trondheim, Norway, August 30 - September 2, 2005. ACM 2005
- Mario A. Nascimento, M. Tamer zsu, Donald Kossmann, Rene J. Miller, Jos A. Blakeley, K. Bernhard Schiefer (Eds.): (e)Proceedings of the Thirtieth International Conference on Very Large Data Bases, Toronto, Canada, August 31 - September 3 2004. Morgan Kaufmann 2004
- Peter M. G. Apers, Paolo Atzeni, Stefano Ceri, Stefano Paraboschi, Kotagiri Ramamohanarao, Richard T. Snodgrass (Eds.): VLDB 2001, Proceedings of 27th International Conference on Very Large Data Bases, September 11-14, 2001, Roma, Italy. Morgan Kaufmann 2001
- Amr El Abbadi, Michael L. Brodie, Sharma Chakravarthy, Umeshwar Dayal, Nabil Kamel, Gunter Schlageter, Kyu-Young Whang (Eds.): VLDB 2000, Proceedings of 26th International Conference on Very Large Data Bases, September 10-14, 2000, Cairo, Egypt.
- Jorge B. Bocca, Matthias Jarke, Carlo Zaniolo (Eds.): Proceedings of the 20th International Conference on Very Large Data Bases, (VLDB'94), September 12-15, 1994, Santiago de Chile, Chile. Morgan Kaufmann
- IEEE/WIC/ACM International Conference on Web Intelligence, WI 2009, Milan, Italy, 15-18 September 2009, Main Conference Proceedings. IEEE 2009
- IEEE / WIC / ACM International Conference on Web Intelligence, WI 2008, 9-12 December 2008, Sydney, NSW, Australia, Main Conference Proceedings. IEEE 2008
- IEEE / WIC / ACM International Conference on Web Intelligence (WI 2006), 18-22 December 2006, Hong Kong, China. IEEE Computer Society 2006
- IEEE / WIC International Conference on Web Intelligence, (WI 2003), 13-17 October 2003, Halifax, Canada. IEEE Computer Society 2003
- Manuela M. Veloso (Ed.): IJCAI 2007, Proceedings of the 20th International Joint Conference on Artificial Intelligence, Hyderabad, India, January 6-12, 2007
- Georg Gottlob, Toby Walsh (Eds.): IJCAI-03, Proceedings of the Eighteenth International Joint Conference on Artificial Intelligence, Acapulco, Mexico, August 9-15, 2003. Morgan Kaufmann 2003
- Abraham Bernstein, David R. Karger, Tom Heath, Lee Feigenbaum, Diana Maynard, Enrico Motta, Krishnaprasad Thirunarayan (Eds.): The Semantic Web - ISWC 2009, 8th International Semantic Web Conference, ISWC 2009, Chantilly, VA, USA, October 25-29, 2009. Proceedings. Lecture Notes in Computer Science 5823 Springer 2009

- Amit P. Sheth, Steffen Staab, Mike Dean, Massimo Paolucci, Diana Maynard, Timothy W. Finin, Krishnaprasad Thirunarayan (Eds.): The Semantic Web - ISWC 2008, 7th International Semantic Web Conference, ISWC 2008, Karlsruhe, Germany, October 26-30, 2008. Proceedings. Lecture Notes in Computer Science 5318 Springer 2008
- Isabel F. Cruz, Stefan Decker, Dean Allemang, Chris Preist, Daniel Schwabe, Peter Mika, Michael Uschold, Lora Aroyo (Eds.): The Semantic Web - ISWC 2006, 5th International Semantic Web Conference, ISWC 2006, Athens, GA, USA, November 5-9, 2006, Proceedings. Lecture Notes in Computer Science 4273 Springer 2006
- Sheila A. McIlraith, Dimitris Plexousakis, Frank van Harmelen (Eds.): The Semantic Web - ISWC 2004: Third International Semantic Web Conference, Hiroshima, Japan, November 7-11, 2004. Proceedings. Lecture Notes in Computer Science 3298 Springer 2004
- Isabel F. Cruz, Vipul Kashyap, Stefan Decker, Rainer Eckstein (Eds.): Proceedings of SWDB'03, The first International Workshop on Semantic Web and Databases, Co-located with VLDB 2003, Humboldt-Universitt, Berlin, Germany
- Wessel Kraaij, Arjen P. de Vries, Charles L. A. Clarke, Norbert Fuhr, Noriko Kando (Eds.): SIGIR 2007: Proceedings of the 30th Annual International ACM SIGIR Conference on Research and Development in Information Retrieval, Amsterdam, The Netherlands, July 23-27, 2007. ACM 2007
- Efthimis N. Efthimiadis, Susan T. Dumais, David Hawking, Kalervo Jrvelin (Eds.): SIGIR 2006: Proceedings of the 29th Annual International ACM SIGIR Conference on Research and Development in Information Retrieval, Seattle, Washington, USA, August 6-11, 2006. ACM 2006
- SIGIR '98: Proceedings of the 21st Annual International ACM SIGIR Conference on Research and Development in Information Retrieval, August 24-28 1998, Melbourne, Australia. ACM 1998
- James Bailey, David Maier, Klaus-Dieter Schewe, Bernhard Thalheim, Xiaoyang Sean Wang (Eds.): Web Information Systems Engineering - WISE 2008, 9th International Conference, Auckland, New Zealand, September 1-3, 2008. Proceedings
- Jinpeng Huai, Robin Chen, Hsiao-Wuen Hon, Yunhao Liu, Wei-Ying Ma, Andrew Tomkins, Xiaodong Zhang (Eds.): Proceedings of the 17th International Conference on World Wide Web, WWW 2008, Beijing, China, April 21-25, 2008. ACM 2008
- Carey L. Williamson, Mary Ellen Zurko, Peter F. Patel-Schneider, Prashant J. Shenoy (Eds.): Proceedings of the 16th International Conference on World Wide Web, WWW 2007, Banff, Alberta, Canada, May 8-12, 2007. ACM 2007
- Les Carr, David De Roure, Arun Iyengar, Carole A. Goble, Michael Dahlin (Eds.): Proceedings of the 15th international conference on World Wide Web, WWW 2006, Edinburgh, Scotland, UK, May 23-26, 2006. ACM 2006
- Allan Ellis, Tatsuya Hagino (Eds.): Proceedings of the 14th international conference on World Wide Web, WWW 2005, Chiba, Japan, May 10-14, 2005. ACM 2005

- Stuart I. Feldman, Mike Uretsky, Marc Najork, Craig E. Wills (Eds.): Proceedings of the 13th international conference on World Wide Web, WWW 2004, New York, NY, USA, May 17-20, 2004. ACM 2004
- International World Wide Web Conferences Steering Committee (IW3C2), Proceedings of the Twelfth International World Wide Web Conference, WWW2003, Budapest, Hungary, 20-24 May 2003. ACM 2003
- International World Wide Web Conferences Steering Committee (IW3C2), Proceedings of the Tenth International World Wide Web Conference, WWW 10, Hong Kong, China, May 1-5, 2001. ACM 2001
- Lora Aroyo, Paolo Traverso, Fabio Ciravegna, Philipp Cimiano, Tom Heath, Eero Hyvnen, Riichiro Mizoguchi, Eyal Oren, Marta Sabou, Elena Paslaru Bontas Simperl (Eds.): The Semantic Web: Research and Applications, 6th European Semantic Web Conference, ESWC 2009, Heraklion, Crete, Greece, May 31-June 4, 2009, Proceedings. Lecture Notes in Computer Science 5554 Springer 2009
- John F. Elder IV, Franoise Fogelman-Souli, Peter A. Flach, Mohammed Javeed Zaki (Eds.): Proceedings of the 15th ACM SIGKDD International Conference on Knowledge Discovery and Data Mining, Paris, France, June 28 - July 1, 2009. ACM 2009
- Ying Li, Bing Liu, Sunita Sarawagi (Eds.): Proceedings of the 14th ACM SIGKDD International Conference on Knowledge Discovery and Data Mining, Las Vegas, Nevada, USA, August 24-27, 2008. ACM 2008
- Pavel Berkhin, Rich Caruana, Xindong Wu (Eds.): Proceedings of the 13th ACM SIGKDD International Conference on Knowledge Discovery and Data Mining, San Jose, California, USA, August 12-15, 2007. ACM 2007
- Tina Eliassi-Rad, Lyle H. Ungar, Mark Craven, Dimitrios Gunopulos (Eds.): Proceedings of the Twelfth ACM SIGKDD International Conference on Knowledge Discovery and Data Mining, Philadelphia, PA, USA, August 20-23, 2006. ACM 2006
- Won Kim, Ron Kohavi, Johannes Gehrke, William DuMouchel (Eds.): Proceedings of the Tenth ACM SIGKDD International Conference on Knowledge Discovery and Data Mining, Seattle, Washington, USA, August 22-25, 2004. ACM 2004
- Lise Getoor, Ted E. Senator, Pedro Domingos, Christos Faloutsos (Eds.): Proceedings of the Ninth ACM SIGKDD International Conference on Knowledge Discovery and Data Mining, Washington, DC, USA, August 24 - 27, 2003. ACM 2003
- Proceedings of the sixth ACM SIGKDD international conference on Knowledge discovery and data mining, August 20-23, 2000, Boston, MA, USA. ACM 2000
- Thanaruk Theeramunkong, Boonserm Kijsirikul, Nick Cercone, Tu Bao Ho (Eds.): Advances in Knowledge Discovery and Data Mining, 13th Pacific-Asia Conference, PAKDD 2009, Bangkok, Thailand, April 27-30, 2009, Proceedings. Lecture Notes in Computer Science 5476 Springer 2009
- Proceedings of the 3rd IEEE International Conference on Semantic Computing (ICSC 2009), 14-16 September 2009, Berkeley, CA, USA. IEEE Computer Society 2009

- Proceedings of the First SIAM International Conference on Data Mining, April 5-7, 2001, Chicaco, Illinois, USA. SIAM 2001
- Gerhard Weikum, Arnd Christian König, Stefan Deßloch (Eds.): Proceedings of the ACM SIGMOD International Conference on Management of Data, Paris, France, June 13-18, 2004. ACM 2004
- 6th Atlantic Web Intelligence Conference, September 9-11, 2009 - Prague, Czech Republic
- Jean-Franois Boulicaut, Floriana Esposito, Fosca Giannotti, Dino Pedreschi (Eds.): Knowledge Discovery in Databases: PKDD 2004, 8th European Conference on Principles and Practice of Knowledge Discovery in Databases, Pisa, Italy, September 20-24, 2004, Proceedings. Lecture Notes in Computer Science 3202 Springer 2004
- Zoubin Ghahramani (Ed.): Machine Learning, Proceedings of the Twenty-Fourth International Conference (ICML 2007), Corvalis, Oregon, USA, June 20-24, 2007. ACM International Conference Proceeding Series 227 ACM 2007
- Carla E. Brodley, Andrea Pohoreckyj Danyluk (Eds.): Proceedings of the Eighteenth International Conference on Machine Learning (ICML 2001), Williams College, Williamstown, MA, USA, June 28 - July 1, 2001. Morgan Kaufmann 2001
- Chee Yong Chan, Prasenjit Mitra (Eds.): 11th ACM International Workshop on Web Information and Data Management (WIDM 2009), Hong Kong, China, November 2, 2009. ACM 2009
- Roger H. L. Chiang, Alberto H. F. Laender, Ee-Peng Lim (Eds.): Fifth ACM CIKM International Workshop on Web Information and Data Management (WIDM 2003), New Orleans, Louisiana, USA, November 7-8, 2003. ACM 2003
- Roger H. L. Chiang, Ee-Peng Lim (Eds.): 3rd International Workshop on Web Information and Data Management (WIDM 2001), Friday, 9 November 2001, In Conjunction with ACM CIKM 2001, Doubletree Hotel Atlanta-Buckhead, Atlanta, Georgia, USA. ACM, 2001

1.5.4   Books

- Baeza-Yates, R. and Ribeiro-Neto, B. *Modern Information Retrieval*. Addison-Wesley, 1999. Second edition will apear in 2010.
- Euzenat, J. and Shvaiko, P. *Ontology Matching*. Springer-Verlag, Berlin Heidelberg (DE), 2007.
- Liu, B. (Ed.). *Web Data Mining: Exploring Hyperlinks, Content and Usage Data*. Springer Berlin-Heidelberg, 2006.
- Velásquez, J. D. and Palade, V. *Adaptive Web Sites: A knowledge extraction from web data approach*. IOS Press, Amsterdam, NL, 2008.
- Hastie, T., Tibshirani, R., and Friedman, J.. *The Elements of Statistical Learning: Data Mining, Inference, and Prediction, Second Edition (Springer Series in Statistics)*. Springer-Verlag, 2nd ed. 2009.
- Inmon, W. H. *Building the Data Warehouse, 4rd Edition*. Wiley Publishing, 2005.

- Kimball, R. and Ross, M. *The Data Warehouse Toolkit: The Complete Guide to Dimensional Modeling (Second Edition)*. Wiley, 2002.
- Kohonen, T., Schroeder, M. R., and Huang, T. S. (Eds). *Self-Organizing Maps*. Springer-Verlag New York, Inc., Secaucus, NJ, USA, 2001.
- Markov, Z. and Larose, D. T. *Data Mining the Web: Uncovering Patterns in Web Content, Structure, and Usage*. Wiley-Interscience, 2007.
- Mitchell, T. M. *Machine Learning*. McGraw-Hill, New York, 1997.
- Schölkopf, B. and Smola, A.J. *Learning with Kernels: Support Vector Machines, Regularization, Optimization, and Beyond*. MIT Press, Cambridge, MA, USA, 2001.
- Vapnik, V. N. *The Nature of Statistical Learning Theory (Information Science and Statistics)*. Springer, 1999.
- Graham, L. *A pattern language for web usability*. Addison-Wesley, 2003.
- Han, J., Kamber, M. *Data mining: Concepts and Techniques*, Morgan Kaufmann Publishers Inc., San Francisco, CA, 2000.
- Dorogovtsev, S.N., Mendes, J.F.F. *Evolution of Networks: From Biological Nets to the Internet and WWW (Physics)*. Oxford University Press, Inc., New York, NY, USA, 2003.
- Wasserman, S., Faust, K., Iacobucci, D. *Social Network Analysis : Methods and Applications (Structural Analysis in the Social Sciences)*. Cambridge University Press, 1994
- Salomon, D. *Variable-length Codes for Data Compression*. Springer-Verlag New York, Inc., Secaucus, NJ, USA, 2007.
- Ingwersen, P. and Jirvelin, K. *The Turn: Integration of Information Seeking and Retrieval in Context*. Springer, first edition, 2005.
- Kaushik, A. *Web Analytics 2.0: The Art of Online Accountability and Science of Customer Centricity*. Sybex, 2009.
- Langford, D. *Internet ethics*. MacMillan Press Ltd, 2000.
- Manning, C. D. and Schutze, H. *Fundation of Statistical Natural Language Processing*. The MIT Press, 1999.
- Resnick, S. I. *Adventures in stochastic processes*. Birkhauser Verlag, Basel, Switzerland, Switzerland, 1992.
- Wenger, E., McDermott, R., and Snyder, W. *Cultivating communities of practice: A guide to managing knowledge*. Harvard Business School Press, 2002.
- Henninger, M., The Hidden Web, Second Edition, University of New South Wales Press Ltd, Australia, 2008.
- Jain, L.C., Sato, M., Virvou, M., Tsihrintzis, G., Balas, V. and Abeynayake, C. (Eds), *Computational Intelligence Paradigms: Volume 1 – Innovative Applications*, Springer-Verlag, 2008.
- Fulcher, J. and Jain, L.C., *Computational Intelligence: A Compendium*, Springer-Verlag, 2008.
- Virvou, M. and Jain, L.C. (Eds.), *Intelligent Interactive Systems in Knowledge-Based Environments*, Springer-Verlag, 2008.
- Sato, M. and Jain, L.C., *Innovations in Fuzzy Clustering*, Springer-Verlag, 2006.

- Holmes, D. and Jain, L.C. (Eds.), *Innovations in Machine Learning*, Springer-Verlag, 2006.
- Ghosh, A. and Jain, L.C.(Eds.), *Evolutionary Computation in Data Mining*, Springer-Verlag, Germany, 2005.
- Pal, N. and Jain, L.C. (Eds.), *Advanced Techniques in Knowledge Discovery and Data Mining*, Springer-Verlag, London, 2005
- Nikravesh, M., et al. (Ed.), *Enhancing the power of Internet, Springer-Verlag*, Germany, 2004.
- Fulcher, J. and Jain, L.C. (Eds.), *Applied Intelligent Systems*, Springer-Verlag, Germany, 2004.
- Resconi, G. and Jain, L.C., (Eds.) *Intelligent Agents: Theory and Applications*, Springer-Verlag, Germany, 2004.
- Abraham, A. et al. (Ed.), *Recent Advances in Intelligent Paradigms and Applications*, Springer-Verlag, Germany, 2003.
- Howlett, R., Ichalkaranje, N., Jain, L.C. and Tonfoni, G. (Eds), *Internet-Based Intelligent Information Processing*, World Scientific Publishing Company Singapore, 2002.
- Seiffert, U. and Jain, L.C. (Eds.), *Self-Organising neural Networks*, Springer-Verlag, Germany, 2002.
- Jain, L.C., et al. (Eds.), *Intelligent Agents and Their Applications*, Springer-Verlag, Germany, 2002.
- Jain, L.C. and De Wilde, P. (Eds.), *Practical Applications of Computational Intelligence Techniques*, Kluwer Academic Publishers, USA, 2001.
- Jain, L.C. and Fanelli, A.M. (Eds.), *Recent Advances in Artificial Neural Networks: Design and Applications*, CRC Press, USA, 2000.
- Lazzerini, B., et al., *Fuzzy Sets and their Applications to Clustering and Training*, CRC Press USA, 2000.
- Jain, L.C. and Martin, N.M. (Eds.), *Fusion of Neural Networks, Fuzzy Logic and Evolutionary Computing and their Applications*, CRC Press USA, 1999.
- Jain, L.C. and Vemuri, R. (Eds.), *Industrial Applications of Neural Networks*, CRC Press USA, 1998.
- Sato, M. et al., *Fuzzy Clustering Models and Applications*, Springer-Verlag, Germany, 1997.
- Vazirgiannis, M., et al., *Uncertainty Handling and Quality Assessment in Data Mining*, Springer-Verlag, London, 2003.
- Gomez-Perez, et al., *Ontological Engineering*, Springer-Verlag, London, 2004.
- Zhang, S., et. al., *Knowledge Discovery in Multiple Databases*, Springer-Verlag, London, 2004.
- Ko, C.C., *Creating Web-based Laboratories*, Springer-Verlag, London, 2004.
- Grana, M., et al.(Eds.), *Information Processing with Evolutionary Algorithms*, Springer-Verlag, London, 2005.
- Stuckenschmidt, H. and Harmelen, F.V., *Information Sharing on the Semantic Web*, Springer-Verlag, London, 2005.
- Wang, L. and Fu, X., *Data Mining with Computational Intelligence*, Springer-Verlag, London, 2005.

- Abraham, A., Koppen, M. and Franke, K. (Eds.), *Design and Applications of Hybrid Intelligent Systems*, IOS Press, The Netherlands
- Turchetti, C., *Stochastic Models of Neural Networks*, IOS Press, The Netherlands.
- Loia, V. (Editor), *Soft Computing Agents*, IOS Press, The Netherlands.
- Abraham, A., et al. (Eds.), *Soft Computing Systems*, IOS Press, The Netherlands.
- Motoda, H., *Active Mining*, IOS Press, The Netherlands.
- Nayak, R., Ichalkaranje, N. and Jain, L.C. (Editors), Evolution of the Web in Artificial Intelligence Environments, Springer-Verlag, 2008.
- Castellano, G.; Jain, L.C. and Fanelli, A.M. (Editors), Web Personalization in Intelligent Environments, Springer-Verlag, Germany, 2009.
- Lim, C.P., Jain, L.C. and Satchidananda, D. (Editors), Innovations in Swarm Intelligence, Springer-Velag, Germany, 2009.
- Teodorescu, H.N., Watada, J. and Jain, L.C. (Editors), Intelligent Systems and Technologies, Springer-Verlag, Germany, 2009.
- Mumford, C. and Jain, L.C. (Editors), Computational Intelligence: Collaboration, Fusion and Emergence, Springer-Verlag, 2009.
- Nguyen, N.T. and Jain, L.C. (Editors), Intelligent Agents in the Evolution of Web and Applications, Springer-Verlag, Germany, 2009.
- Bianchini, M., Maggini, M., Scarselli, F. and Jain, L.C. (Editors), Innovations in Neural Information Processing Paradigms and Applications, Springer-Verlag, 2010.

References

1. Agrawal, R., Srikant, R.: Privacy-preserving data mining. SIGMOD Rec. 29(2), 439–450 (2000)
2. Baeza-Yates, R.A., Ribeiro-Neto, B.: Modern Information Retrieval. Addison-Wesley Longman Publishing Co., Inc., Boston (1999)
3. Berners-Lee, T., Cailliau, R., Luotonen, A., Nielsen, H.F., Secret, A.: The world wide web. Communications of ACM 37(8), 76–82 (1994)
4. Chair-Giles, C.: Sna-kdd 2009: Proceedings of the 3rd workshop on social network mining and analysis (2009); Program Chair-Giles, C. Lee and Program Chair-Mitra, Prasenjit and Program Chair-Perisic, Igor and Program Chair-Yen, John and Program Chair-Zhang, Haizheng
5. Davis, R., Shrobe, H., Szolovits, P.: What is knowledge representation. AI Magazine 14(1), 17–33 (1993)
6. Dujovne, L.E., Velásquez, J.D.: Design and implementation of a methodology for identifying website keyobjects. In: Velásquez, J.D., Ríos, S.A., Howlett, R.J., Jain, L.C. (eds.) Knowledge-Based and Intelligent Information and Engineering Systems. LNCS, vol. 5711, pp. 301–308. Springer, Heidelberg (2009)
7. Fayyad, U.M., Piatetsky-Shapiro, G., Smyth, P.: From data mining to knowledge discovery: an overview, pp. 1–34. American Association for Artificial Intelligence, Menlo Park (1996)
8. Golbeck, J., Rothstein, M.: Linking social networks on the web with foaf: a semantic web case study. In: AAAI 2008: Proceedings of the 23rd national conference on Artificial intelligence, pp. 1138–1143. AAAI Press, Menlo Park (2008)

9. Grau, B.C., Horrocks, I., Motik, B., Parsia, B., Patel-Schneider, P., Sattler, U.: Owl 2: The next step for owl. Web Semant. 6(4), 309–322 (2008)
10. Han, J., Chang, K.: Data mining for web intelligence. Computer 35(11), 64–70 (2002)
11. Harth, A., Decker, S.: Optimized index structures for querying rdf from the web. In: LA-WEB 2005: Proceedings of the Third Latin American Web Congress, Washington, DC, USA, p. 71. IEEE Computer Society Press, Los Alamitos (2005)
12. Kautz, H., Selman, B., Shah, M.: Referral web: combining social networks and collaborative filtering. Commun. ACM 40(3), 63–65 (1997)
13. Kleinberg, J.M.: Authoritative sources in a hyperlinked environment. J. ACM 46(5), 604–632 (1999)
14. Kosala, R., Blockeel, H.: Web mining research: a survey. SIGKDD Explor. Newsl. 2(1), 1–15 (2000)
15. Kudelka, M., Snásel, V., Horak, Z., Abraham, A.: Social aspects of web page contents. In: Abraham, A., Snásel, V., Wegrzyn-Wolska, K. (eds.) CASoN, pp. 80–87. IEEE Computer Society, Los Alamitos (2009)
16. Liu, B.: Web Data Mining: Exploring Hyperlinks, Content and Usage Data, 1st edn. Springer, Heidelberg (2007)
17. Mika, P.: Social networks and the semantic web. In: IEEE / WIC / ACM International Conference on Web Intelligence, vol. 0, pp. 285–291 (2004)
18. Pechter, R.: What's pmml and what's new in pmml 4.0? SIGKDD Explor. Newsl. 11(1), 19–25 (2009)
19. Rebolledo, V.L., Velásquez, J.D.: A platform for extracting and storing web data. In: Velásquez, J.D., Ríos, S.A., Howlett, R.J., Jain, L.C. (eds.) Knowledge-Based and Intelligent Information and Engineering Systems. LNCS (LNAI), vol. 5711, pp. 843–850. Springer, Heidelberg (2009)
20. Ríos, S.A., Velásquez, J.D.: Semantic web usage mining by a concept-based approach for off-line web site enhancements. In: Web Intelligence, pp. 234–241. IEEE, Los Alamitos (2008)
21. Ríos, S.A., Velásquez, J.D., Vera, E.S., Yasuda, H., Aoki, T.: Improving web site content using a concept-based knowledge discovery process. In: Web Intelligence, pp. 361–365. IEEE Computer Society, Los Alamitos (2006)
22. Shadbolt, N., Berners-Lee, T., Hall, W.: The semantic web revisited. IEEE Intelligent Systems 21(3), 96–101 (2006)
23. Srivastava, J., Cooley, R., Deshpande, M., Tan, P.-N.: Web usage mining: discovery and applications of usage patterns from web data. SIGKDD Explor. Newsl. 1(2), 12–23 (2000)
24. Velasquez, J.D., Palade, V.: Adaptive Web Sites: A Knowledge Extraction from Web Data Approach. IOS Press, Amsterdam (2008)
25. Velásquez, J.D., Palade, V.: A knowledge base for the maintenance of knowledge extracted from web data. Knowledge Based Systems 20(3), 238–248 (2007)
26. Xu, Y., Wang, K., Zhang, B., Chen, Z.: Privacy-enhancing personalized web search. In: WWW 2007: Proceedings of the 16th international conference on World Wide Web, pp. 591–600. ACM Press, New York (2007)
27. Yao, J., Raghavan, V.V., Wu, Z.: Web information fusion: A review of the state of the art. Inf. Fusion 9(4), 446–449 (2008)

Chapter 2
Advanced Techniques in Web Data Pre-processing and Cleaning

Pablo E. Román*, Robert F. Dell, and Juan D. Velásquez

Abstract. Central to successful e-business is the construction of web sites that attract users, capture user preferences, and entice them into making a purchase. Web mining is diverse data mining applied to categorize both the content and structure of web sites with the goal of aiding e-business. Web mining requires knowledge of the web site structure (hyperlink graph), the web content (vector model) and user sessions (the sequence of pages visited by each user to a site). Much of the data for web mining can be noisy. The origin of the noise comes from many sources, for example, undocumented changes to the web site structure and content, a different understanding of the text and media semantic, and web logs without individual user identification. There may not be any record of the number of times a specific page has been visited in a session as page is stored on a proxy or web browser cache. Such noise presents a challenge for web mining. This chapter presents issues with and approaches for cleaning web data in preparation for web mining analysis.

2.1   Introduction

The Web has become the primary communication channel for many financial, trading, and academic organizations. Large corporations such as Amazon and Google would not exist without the Web and they rely on web data as an important source of customer information. For more than 10 years, there have been numerous

Robert F. Dell
Naval Postgraduate School, Operations Research Department,
Monterey, California, USA
e-mail: `dell@nps.edu`

Pablo E. Román · Juan D. Velásquez
University of Chile, Department of Industrial Engineering,
Repblica 701, Santiago, Chile
e-mail: `proman@ing.uchile.cl, jvelasqu@dii.uchile.cl`

 * Corresponding author.

J.D. Velásquez and L.C. Jain (Eds.): Advanced Techniques in Web Intelligence – 1, SCI 311, pp. 19–48.
springerlink.com

methods proposed for extracting knowledge from this web data [79, 70], where pre-processing of data is a critical step toward pattern identification.

Soft computing techniques such as Neural Networks, Fuzzy Logic, Bayesian methods, and Genetic Algorithm are commonly applied to web data to help cope with uncertainty and imprecision [70]. However, these soft computing techniques do not always take care of the time dependency and high dimensionality of web data, suggesting that one takes considerable care in avoiding the celebrated term "Garbage in, Garbage out" [40] . Web data can be placed into three categories: Web Structure, Web Content and Web Usage [67].

Web Structure Data corresponds to the hyperlink network description of a web site. This oriented graph in today's web 2.0 [97] becomes more dynamic and depends on a web user's actions [80]. Web 2.0 applications distinguish themselves from previous software because they take full advantage of dynamic pages and encourage social interaction such as in Facebook and Twitter. This structure was traditionally static [16] but new web applications suggest the need for a time dependent graph.

Web Content Data corresponds to the text content and relates to the field of information retrieval. There is a long history of research on both text retrieval and representation spanning more than fifty years. Vector Space Model or Bag of Word Model [68], are traditional representations, based on word frequencies of text. The focus of recent research is on more accurate representations of the semantic of text [101, 41, 49, 46, 90] and on coping with the more dynamic nature of text. Text now changes depending on both the time and circumstances. The semantic of a web page also changes due to multimedia objects. Parsing web pages, automatic interpretation of objects and pre-processing such information is part of ongoing research process [91, 98].

Web Usage Data corresponds to the trail of pages, also called a session, that tracks each web user while browsing a web site. Monitoring a web user's session can be a violation of privacy [22] and is forbidden by law in several countries [61]. For example, installing tracing software in the web user's browser is equivalent to spyware. There are other less intrusive ways of retrieving sessions; the most popular is by using a web log, a file that records each page retrieved from a web site. This data source is anonymous, but simultaneous interactions of web users makes identifying a session more complicated. With respect to the advancement of browser technology, for instance the Opera software, session retrieval has become more complicated as more sophisticated algorithms for pre-loading and buffering pages have been implemented. New studies [25, 26] relate to increasing the accuracy of sessionization.

Pre-processing web data has some well known issues [87]. Parsing problems relate to different version of HTML, DHTML and non-compliant codes. Dynamic content generated via embedded code like JavaScript or server side content generation render text content extraction unfruitful through usual parsing methods. Pages with frames produce different presentations for a user that can be interpreted as a "pageview" [70] that combines group pages and other objects together.

A pageview abstract is a specific type of user activity such as reading news, or browsing search results. Even when a session is considered as a sequence of pageviews, dynamic pages produce a large number of possible object combinations for each pageview. Log files are influenced by new browser technology (linkprefetching) [56], enhanced cache management and also by network caching devices such as proxies and parallel browsing consisting of tabs and popups.

The quality of pre-processing has been quantified using a set of measures for each data type. New measures have also been used in the case where uncertain results are obtained [25]. An efficient storage and data representation is also important because companies like Google have large data storage that incorporates most of the Web. This chapter, aims to explore the difficulty in obtaining quality data. In section 2.2 general statistical characteristics of web data are discussed. Section 2.3 presents the state the art in web graph retrieval. Section 2.4 expounds on the current methodology for retrieving text data. In Section 2.5 the process of session retrieval is analyzed and examples are provided. Finally section 2.6 provides a summary.

2.2   The Nature of the Web Data: General Characteristics and Quality Issues

Web data is not a random collection of users' interactions, pages and content. There is some regularity that remains constant across the sites and groups of web user sessions (e.g., [42]). This should impact pre-processing [25] and the evaluation of web mining.

Knowing prior web data regularities could significantly improve data mining. For example, for pattern recognition, semi-supervised clustering techniques have been popular in recent years. The results found by these algorithm are considerably better [75] because they use domain data descriptions that refine the search space where an unsupervised machine learning algorithm works [37]. The resulting subspace region is less of a factor with the additional domain information. In the field of data cleaning, deviations identify outliers or automatic anomaly detection [69].

2.2.1   Web Content

Web content information retrieval has been studied for many years. For example, the evaluation of a text for word frequencies. Today, Web content is based on mul-timedia in which rich text features enhance a web user's experience. In this context, extracting the semantics relies on the discovery of compact structures or a web object that represents a component in a web page.

Pure web texts do not differ from a traditional corpus and empirical studies reveal statistical regularities on the text distributions. The well known Zipf Law from 1932 [45, 68] states that in a corpus the ranked number of words per page is the power $\alpha \simeq 1$ (see equation 2.1) . Today, linguistics agree that this rule assumes speakers simplify communication by using a small pool of words that can be retrieved efficiently from their memory. Furthermore, the listeners simplify communication by selecting words with a single unambiguous meaning. Other models relate to a stochastic diffusion model for generalizing Zipf's Law [58] and cast a new perspective on the subject.

$$P(n) = \frac{n^{-\alpha}}{\sum_{k=1}^{\infty} k^{-\alpha}} \tag{2.1}$$

One of most celebrated distribution is the Heaps' Law [63] used in recent applications to describe the Internet. The heap law describes the number of unique word in a text as a power with exponent β representing its size or the number of word on a page. It has been found that $\beta = 0.52$ [62] has greater accuracy on internet pages. The study then estimates the number of n-tuples of words m having the expected number of hits on search engines, obtaining log-normal results. Also these distributions are used to measure similarity between pairs of words for grouping purposes. This kind of clustering result is useful for grouping terms on texts that capture the semantic for further processing more effectively.

Thanks to the advent of Web 2.0 applications , the content of web pages has become more dynamic. Updates on web sites have been a prolific subject of focus since web search engines are based on the accuracy of indexed terms and pages. The study [34] includes 151 millions Web pages browsed once a week within a three month period. The findings signified that 22% of pages were deleted, and 34.8% of pages had content updates (larger pages where more frequently updated). From other studies [78] the information longevity is not related to the extent of updates on the web page. It also reveals that dynamic changes can be placed into two main categories: Churn Updating behavior (33%) related to repeatedly overwriting content and Scroll Updating behavior (67%) related to lists of updates (e.g., news).

In a recent study [1] over 55,000 web pages with a diversity of visitation pattern indicated a higher dynamic content than previous studies. Consequently, further refinement of the web content representation and pre-processing must be taken into consideration in order to manage dynamic content. Some other studies reveal that in a one hour period 26% of the visited pages during the study were modified or updated in some way and 66% of those pages were visited the following day, of which 44% consisted of query parameters. Earlier studies, (e.g., [18]) report that only 23% of the pages consisted of dynamic content for the duration of one day.

2.2.2   Web Site Structure

Early studies of web site structure suggested a static graph [16] with pages as nodes and hyperlinks as edges. However, the link structure is as dynamic as content and develops exponentially [6]. Despite the changing structure, large-scale statistical analyses of the hyperlinks [5, 53] show a power law distribution ($p(x) \sim x^{-\alpha}$) has a good fit for several structural measures (categorized by the exponent α). The study [5] suggests that the number of pages per web site is a power law with exponent $\alpha \in [1.2, 1.6]$ [31]; the number of pages that have a given number of in-links have $\alpha \in [1.6, 2.1]$, the number of pages with a given number of out-links has a piecewise power law with $\alpha_1 \in [0.3, 0.7]$ and $\alpha_2 \in [1.9, 3.6]$. Finally the ranking number (Page

Rank) has $\alpha \in [1.8, 1.9]$. The distribution of the age of a page corresponding to the *last modified* value register is a Poisson process [19] with decay parameter $\lambda \in [1.6, 2.3]$.

An important observation taken from current studies is that the notion of a page is losing its importance as a fundamental information source in relation to its description of the overall web structure [6]. In todays dynamic web applications what is commonly described as a web page depends on the parameters that the application receives. Today, a complete web site could be served by a unique application file in which each page is represented by the URL's query parameters.

As stated by [30], the temporal component of the graph evolution has not been conclusively studied by the web mining community. They consider three levels of study: temporal "single nodes", temporal "sub-graphs" and "whole graph" analysis. The single node study examines how frequently a page is accessed during a time period in which there are no changes to a page. In this case, the data mining approach consists of clustering over page content. The subgraph level consists of finding the period of time where a small level of change has occurred within a structure based on graph properties such as order, size, components, "Max Authorities", "Hub", and "Page Rank". The "whole graph" analysis focuses on identifying a set of measures for building a features vector of the graph. Like in the subgraph level, the features could be represented by basic graph properties (e.g., order and size) or derived properties (e.g., "Max Hub" score and "Average Hub" score). A current means of research is to explain the changes of such measures in time.

2.2.3   Web User Session

The celebrated Law of Surfing [42, 104, 43] was noticed ten years ago as a distribution that regularly fits the number of sessions of a given size. This distribution was recognized as an inverse Gaussian (see equation 2.2) whose parameters change depending on the Web site ($E[L] = \mu, Var[L] = \mu^3/\lambda$). This law is also approximated by the Zipf's laws simplifying the calculation process as in log scale the fitting problem reduce to linear regression.

$$P(L) = \sqrt{\frac{\lambda}{2\pi L^3}} e^{-\lambda(L-\mu)^2/2\mu^2 L} \qquad (2.2)$$

Some empirical studies have been performed on browsing customs [95, 13, 102], with recent changes on web user general behavior reported in [103]. The change in browsing habits can be explained by the highly dynamics environments and new internet services. The use of a browser's interface widget orchestrates the changes in frequencies. For example users backtracking on certain sites are declining. Earlier studies [106] report 30% back button activation. However, new studies [77] report only 14% revisitation patterns. This behavior is most likely also the outcome of new applications incorporating backtracking support within a web page. The use of the forward button also resulted in decrease usage; decreasing from 1.5% to 0.6%. The reload button usage has also decreased from 4.3% to 1.7%.

New windows and submitting activities reported an increase in total user's actions of ($\sim$ 10%), as they are key features of dynamic pages. Despite this, the link click stream seems to maintain its percentage of actions in web browsers at 44%. As reported in [77], search engines, now considered new browsing assisting tools, carried out 16% of page visits. Another new behavior also appears as navigating simultaneously with new windows and tabs, where an average of two pages are opened throughout a session. It confirms that parallel navigation is not the exception but the rule.

The same study [77] shows a heavy tail distribution of the average stay time on a page where a considerable fraction of the visits last a very short time. Revisiting pages or scanning behavior corresponds to a fraction of the entire user's action. Despite these observations, a direct correlation is found between time remaining on the site and the number of hyperlinks on the page (also with the number of words).

[3, 2] perform other empirical work on navigating of electronic documents like pdf and MS Word. Despite the many differences between users, they report some similarities. A power law distribution models the number of times a document opens. Some interesting patterns were discovered by mean of a tracking system on the user interface action. The period of time spent reading a section of the text appears to relate to the interest of the user and to the quality of content. In these cases, the main methods of navigating correspond to scrolling operations, where distances travelled in relation to session size are revealed to have a power law distribution. They report the same behavior on the mean percent of time in text regions with a heavy tail distribution. A general rule for human behavior seeking information seems to rule the interactions on internet and document seeking behavior.

Web usage research is mainly driven by click stream data (web user's sessions). A recent study on more precise biometric data in conjunction with Web Usage Data shows that click streams are a bias source of information [50, 38]. Study [38] uses eye monitoring techniques to focus on the click stream in a Google search result page. This reflects how web users' behave in relation to the texts and links. Eye tracking data provides important information about the web user's cognitive processing that have a direct relation to their actions. This study highlights that click-streams are directly influenced by the order of the presented search result; the first links on a page are likely to be chosen as they tend to be inspected to a greater extent. The first visual portion of the web page also has a direct influence on the click decision as a web user prefers to process this first portion. In this way, sessions obtained from web logs do not directly reflect web user behavior. The visual disposition of the elements on the web page must also be taken into consideration.

2.2.4 Privacy Issues

Legal issues must be considered when collecting web usage data [44]. The extraction of information from web data can lead to the identification of personal data that can be illegal to collect. Any general solution to issues of what to collect must balance four forces: laws, social norms, the market, and technical mechanism.

Privacy relates to "The control of information concerning an individual," "the right to prevent commercial publicity of one's own name and image," "the right to make certain personal and intimate decisions free from government interference," and "to be free from physical invasion of one's home or person" [73]. Certainly, a whole spectrum of solutions exists but all of them need to be tested under current laws and social norms. While one method may protect individuals, it could compromise social information (e.g., racial discrimination) and can lead to legal prosecutions.

Legislation on privacy control is still unclear [73]. The W3C organization produces a policy called the P3P Platform for Privacy Preferences [66]. This policy stipulates that information about web users must be protected by privacy laws and the users themselves informed about their rights. It suggests [73] a continuous process of negotiating, the inclusion of relevant third parties and an optimum or acceptable level of disclosure of personal information in an online environment." Hence, an optimal level of privacy strives to return the privacy control to the web user, and declares that browsing preferences must determine the level of privacy.

A user's privacy can be resolved by incorporating a "slider" on a web browser that controls behavior monitoring. The P3P protocol should be the mechanism that determines and illustrates the level of privacy a user is entitled to in each web site. The use of this protocol is increasing among web sites [85]. For instance, with a common session tracking mechanism (section 2.5.2.1) the P3P protocol defines the following cookie persistent or non-persistent, first party or third party, non-personal or personal. "Persistent" cookies correspond to a permanent cookie repository that can store all interaction within the web. "First party" cookies send information only to the web server of the page while "Third party" allows to distribute data to any other server. "Personal" cookies store any personal web user data. According to this protocol, the most privacy compliant cookie is a non-persistent, first party and non-personal. The worst privacy compliant cookie is a persistent, third party and personal. This information will be present on the P3P protocol and future non-compliant privacy tracking session methods will be banned by most users [73].

2.2.5 *Quality Measures*

Ensuring web data quality is important for data mining. We will be presented in this chapted different quality measures for different types of web data. Quality metrics also enable the implementation of pre-processing filters.

Web Structure Data

The discipline of graph mining [14] relates to the analysis of the web structure's intrinsic graph properties. Data quality depends on the application where the data is used. For example, in web structure mining, large web crawling results need to be evaluated for the page rank value as a measure of the importance of a page [93]. The Page Rank value is an indicator of the visibility that a web page has on a search engine. Another metric corresponds to the hub and authority scores [14] that

concentrates on out-going and in-going links, [84] discusses more indirect measures
based on the web user action on the search results.

$$c_i = \begin{cases} \frac{n_i}{k_i} & k_i > 1 \\ 0 & k_i \in \{0, 1\} \end{cases} \tag{2.3}$$

There are graph theoretical measures related to community structure [14] that are
useful for controlling data quality. The clustering coefficient (Equation 2.3 where k_i
is number of neighbors of node i and n_i the number of edges between them) reflects
the degree of transitivity of a graph. Other measures relate to identifying the number
of connected sub-graph components. Thus a disconnected web graph is an indicator
of some kind of problem in the data collection or a recent change. Another important indicator is the resilience value [14]. This is obtained by finding the minimum
cut-set cardinality of the graph. A cut-set is a set of links that split the graph into
two components. A data set with a high resiliency value should be the most representative.

Web Content Data
Web pages are special cases in the context of text pre-processing, since HTML tags
contain both embedded and miscellaneous information (e.g., advertisement and lo-
gos). It is estimated that an average of 13.2% of the embedded information on web
pages is noise related [111]. A similar value found between vector model text content and a predefined set of topics provides an indication of the quality of the content
data [111]. Filtering embedded information familiar with this measure helps with
pre-processing text data.

Web Usage Data
The set of web users' sessions correspond to web usage data. The quality of such
large data could be tested using the session size distribution, which follows a site independent distribution law [42]. For simplicity, this distribution is approximated by
a power law [64]. The goodness of fit value to this distribution is a quality measure
of the session set [25, 26, 27].

If real sessions are available (for example using logged users), then the "degree
of overlap" between sessions can be compared [94]. It is defined as the maximal
averaged fraction of real sessions that are recovered using the sessionization. Another wide spread measure [7] corresponds to the precision and recall scheme. The
precision p is a measure of exactness that can be defined as $p = N_p/N_s$, where N_p
is the number of maximal patterns positively recovered and N_s is the number of
maximal patterns recovered from the sessionization. The recall measure is a measure of completeness that can be defined as $r = N_p/N_r$, where N_r is the number of
correct maximal patterns. An accuracy value can be defined as a geometric mean
$a = \sqrt{p * r}$. Greater accuracy means a more effective sessionization method. Simulation can also be used to obtain artificial sessions [94, 7]. Such simulations should
be "human compliant" with statistical behavior following known strong regularities [42]. Some such simulation results are available [86].

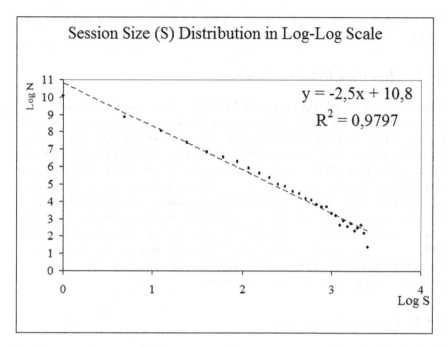

Fig. 2.1 Session size distribution shows an approximate power law for a cookie sessionization. Data extracted April-June 2009 from http://dii.uchile.cl web site

General Measures

Data mining algorithms are influenced by data quality [107]. For instance, a neural network is particularly sensitive to this issue as reported in [108]. One way to handle this is by using anomaly detection algorithms. Because an anomaly is based on parameter settings it can be considered as a metric of data quality. Support vector machines (SVM) have been largely used for these purposes [100]. SVM can learn complex regions and locate outliers. The fraction of outliers provides a measure of the quality of the data. This general measuring principle can be applied to web feature vectors as defined in section 2.5.1. Furthermore, finding the outliers implies a mechanism for data cleaning.

2.3   Transforming Hyperlinks to a Graph Representation

With respect to the advent of search engines, the field of structure retrieval has been widely studied. The web structure as a graph representation has it own specific data mining process [14]. Large scale retrieval schema spanning the Web has been implemented successfully. Web page structure must be retrieved by sampling web pages following the observed hyperlink structure. This process is called Web

Crawling [12] and involves the storage of the hyperlink structure and web page content.

2.3.1 Hyperlink Retrieval Issues

A major difficulty for hyperlink retrieval is large volumes and the frequent rate of change regarding web data [1]. The quantity of data slows the complete retrieval process. A large crawling of the web must select an update strategy such as selecting the pages most likely to change first. The crawler must have a set of policies regarding: the page selection method, the revisit schedule, a politeness policy, and parallel processing. Several strategies are available based on incomplete information: Breadth search ordering, Page rank based, prediction of changing pages [53]. Other issues relate to the imperfect mapping between URL and the page that is visually seen by users [87]. A common HTML structure like frameset, groups a set of URLs in the same virtually presented page. Hence a frameset produces confusion in the process of mapping URLs to web pages. Nevertheless, a solution considers the group of web pages as a "pageview" object [70].

2.3.2 Crawler Processing

The high rate of web page modification [1] suggests that the time of page retrieval could be near the time of page content expiration. It is important to prioritize the most important pages for retrieval, and that requires a measure of importance for the selection algorithm. The revisiting policy should seek to maintain the freshness of the pages, and once predicting an updated page, it is inserted on the pile for crawling (Figure 2.2). The strategy for revisiting pages could be using the same frequency for all (Uniform case), proportional to its registered updating frequency, or futher estimation for the time of page obsolescence. A recent method based on the longevity of web pages [78] optimizes the cost of page retrieval using a generative model. Web pages are modelled as an independent set of content regions of three types: Static, Churn Content, Scroll Content. Such content types have specific lifetime probability distributions that are fitted with observed data. The page lifetime is then estimated using this distribution. The retrieval mechanism must ensure is does not overload the web server. This is called the web crawler's politeness policy, if they are not compliant, the retrieval program could be blocked on the network by a third parties. The visit interval setting for the same web server should have a lower limit (usually 10 seconds) that ensures a minimal impact on the web server. This produces a slow retrieval rate because a web site can have anywhere from a thousand to a million pages. Finally a parallel algorithm reduces the processing time but should be designed to avoid retrieving the same page twice.

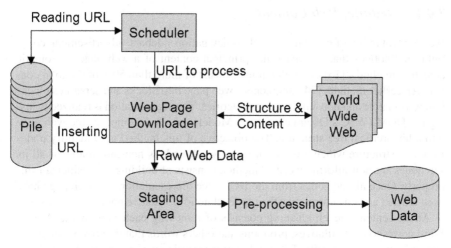

Fig. 2.2 High level Crawler Structure

2.3.3 *Large Sparse Distributed Storage*

A crawler that inserts data directly into a database results in poor performance [12]. Retrieved data is therefore stored on a intermediary format for further processing. Recent advances on storage for distributed systems [17] give some insight into efficiently connecting a web crawler with structured data storage. The Google's Bigtable storage engine is a distributed storage engine for peta byte size. It is implemented over several projects including web indexing. It consists of a three dimensional map of strings that are indexed by row, column and timestamp. A hyperlink structure fits well in this schema because a row can represent the page of a hyperlink, the column can correspond to the pointed page, and the timestamp represents the link when it's retrieved. This storage facility is implemented using the Google File System and operates over a shared pool of servers. It is built on top of the Google SSTable file format that provides a persistent ordered immutable map from keys to values.

2.4 Transforming Web Content into a Feature Vector

The information presented on web pages is complicated by text and multimedia content. This information is used for page classification [83], page summarization [89], entity extraction [98] and semantic processing. This data needs to be filtered because many objects present on web pages are not related with the stated objectives (e.g., advertising). Furthermore text content by itself is noisy, for example sentences contain many word that have a poor influence on its semantic content. Text algorithms are also dependent on the text's language. [111] present a collection of text preprocessing algorithms for data mining purposes.

2.4.1   Cleaning Web Content

Web content is full of noisy items such as navigation sidebars, advertisement, copy-right, and notices that can mask the principal content of a web page. Automatic detection methods, [15] claim that noise consist of more than 50% of the total content. An earlier method, [24], segments a web page into blocks and separates blocks into informative and non-informative categories. The segmentation is realized by using the DOM (W3C's Document Object Model), which decomposes the web page into a hierarchical tree structure. The objective of this method is to identify nodes from this structure which are non-informative. A node is non-informative if all its sub nodes are non-informative. The method consists of building a classifier for this property. Once all the nodes from the DOM tree representation of a web page have been labeled as non-informative the top level non-informative nodes are removed.

At the sentence level, cleaning consists of removing each Stop Word. A Stop Word corresponds to illatives, pronouns and others words that do not contain much semantic value individually. This kind of preprocessing is fundamental for the "Bag of Word" model where the semantic is approximated by a set of words without considering the sentence and paragraph structure semantic. An additional step is known as Porter Stemming that reduces each word to it root [82].

In English, the algorithm for stemming is simple because it corresponds to identifying the suffix and removing it. In other languages, it can be more complicated so specific stemming algorithms in several languages exist [92]. Another approach for stemming sentences consists of clustering words based on a corpus statistic [8]. The algorithm is trained on a corpus that finally defines a representative word from a set of words, the stemmed text has been shown to produce better results for automatic classification.

The changing content of a web page [1] must be managed using time window processing [47] where it is assumed that content remains constant during a defined period of time. This implies a content feature vector update using a temporal index.

A promising direction on text pre-processing is the Word Sense Disambiguation technique [96] (WSD). It addresses the problem of selecting the most appropriate sense for a word with respect to its context. The technique consists of selecting the most appropriate meaning for each word using a semantic model of the text. It was reported that WSD boosted the further information retrieval and data mining processing [105].

2.4.2   Vector Representation of Content

The simplest text representation consist of a vector $V = [\omega_i]$ of real components indexed by word. This representation comes from the Bag of Word abstraction. Despite the approximation, this model provides remarkable positive results in the information retrieval [68]. The value of each component of the vector ω_i is called a weight for word i. There are several weighting schema for words (Term), the most common of which is the TF-IDF schema. Table 2.1 presents the most common weightings.

Table 2.1 Common weighting Schema. Index i is for a unique term, index k for a unique document. A document k is represented by a vector $[\omega_{ik}]_{i=1,\ldots,N_k}$, where $\omega_{ik} = f(\textit{term } i \textit{ in document } k)$. n_{ik} is the times the term i appears in the document k and n_{ik}^- correspond to the negative number of appearances predicted for the word i according to a trained algorithm. N_k is the number of terms in documents k

$f(.)$	Calculation
binary	$binary(\omega_i) = 1$ if the term i is present on the document k or 0 if not.
tf_k	Term frequency on the document k, $tf_k = n_{ik}/N_k$, n_{ik}/N_k is the frequency of the term i on the document k.
$logtf_k$	$logtf_k = log(1 + tf_k)$
ITF	Inverse Term Frequency, $ITF = 1 - 1/(1 + tf)$
idf	Inverse Document Frequency, $idf = log(N/n_i)$, where n_i is the number of document having the term ω_i and N the total number of document.
$tf.idf$	$tf.idf = tf * idf$ [68]
$logtf.idf$	$logtf.idf = log(1 + tf)idf$
$tf.idf - prob$	The probabilistic approximation of the value $tf.idf$ using an estimator for idf
$tf.chi$	Use of the χ^2 feature selection measure, $tf.chi = tf * \chi^2$
$tf.rf$	$tf.rf = tf * log(1 + n_{ik}/max\{1, n_{ik}^-\})$, where rf is the relevance frequency [59, 60]
$tf.ig$	$tf.ig = tf * ig$, where ig is the information gain (Kullback Leibler divergence [33]).
$tf.gr$	$tf.gr = tf * gr$, where gr is the information gain ratio [74].
$tf.OR$	$tf.OR = tf * OR$, where OR is the Odds Ratio [36].

The replacement of a document by a "Bag of Word" inevitably involves a loss of information. For instance, "The Matrix" and "Matrix" represent a film and a mathematical term. Ontology gives a more correct description of the semantic of objects on a web page. Once the ontology annotation on web objects is complete it is possible to transform it into a vector representation [11]. But first an automatic semantic annotation should be performed [81].

2.4.3 Web Object Extraction

Extracting information automatically or semiautomatically from web data has become more difficult as web sites have adopted multimedia technologies to enhance both their content and their presentation. Some of the most successful sites provide video streaming and picture sharing. Unlike text, the content of multimedia formats within web pages can be understandable only to humans. At the most, some technical information such as the color of a histogram or wave patterns for pictures and sounds can be obtained automatically. Given this, a different approach to data extraction for these formats must be adopted.

The use of metadata to describe the content of any multimedia format allows the creation of automatic or semiautomatic ways of extracting information. This enables the webpage to be described as a series of objects brought together in an ordered manner similar to a structured manner in which text and multimedia formats are displayed within a page.

An object displayed within a webpage is termed a Web Object. One definition is found in [32], a web object corresponds to text and multimedia content that a user identifies as a compact unity. Using this definition every part of a webpage can be describe as an object, a picture a sound or even a paragraph. An advantage here is that the content of a website is described not by the site itself, but in the metadata used to define the Web Objects within it.

Another significant advantage of using Web Objects is that any two objects can be easily compared. This can be achieved by defining a similarity measure that uses the metadata that describes the content of an object. This enables complex comparisons between objects that do not require the same format. For example if a webpage contains only one picture and the accompanying text describes the picture in detail, the metadata for both the picture and the text focus on the content rather than on the format. Therefore by using a similarity measure both objects can be discovered as equivalent.

The development of Web Object techniques is focused mainly through the user's point of view, as they are able to describe a website taking into consideration both the content and appearance of a web page rather than only the data which it contains. Different ways have been developed to describe web pages based on how the user perceives a particular page [10].

A large degree of research has recently been carried out in the field of Web Objects; in this section some of these are described focusing in mainly four areas: Web site Key object identification [32], Web Page Element Classification [10], Named Objects [91] and Entity extraction from the Web.

Web site Key objects Identification: In this work [32], Web Objects are defined using a specially created metadata model and a similarity measure to compare two objects. Data mining reconstructs the user sessions from the web server log and objects are created from the web-pages of a site. By inspecting the users' content preferences and similar sessions (clustering), website key objects can be found which reflect the objects within a site that captivate a user's attention.

Web Page Element Classification: This work [10] creates a method for detecting the interesting areas in a webpage. This is accomplished by dividing a webpage in visual blocks and detecting the purpose of each block based on their visual features.

Named Objects: The manner in which users' interpret a web page lies in the focus of this work [91] where a user's perception of a web page is obtained through the user's intention. Web Design Patterns are then selected based on a user's intentions. These named objects are used as the basis of mining methods which enables Web Content Mining.

Entity Extraction from the Web: A Web knowledge extraction system is created in this work [28, 35], which uses Concepts, Attributes and Entities as input data. By modelling this using ontology, facts from generic structures and formats are extracted. Subsequently a self supervised learning algorithm automatically estimates the precision of these structures.

2.5   Web Session Reconstruction

Sessionization is the process of retrieving web user sessions from web data. Web usage mining highly depends on the correct extraction of Web User sessions [23]. Several methods exist and can be classified according to the level of web user personal data (privacy protection).

Proactive methods directly retrieve a rich set of information concerning the operation of a web user. Examples are cookies based sessionization methods where personal data and activities are stored and retrieved. Other proactive methods consist of installing a tracking application on a user's computer that enables the capture of each interaction with the browser interface. In these cases, privacy issues are raised, and in some countries it is forbidden by law. A further possibility lies in the use of login information to track the web user's actions. In this case, a disclaimer agreement is required between the company and the web user in order to enable the tracking of personal information.

Reactive sessionization corresponds to indirect ways to obtain anonymous sessions. The primary source of data for reactive sessionization is the server Web Logs which contains all activities of all web users excluding personal identifiers. Several heuristics have been used to reconstruct sessions from web logs as individuals can not be uniquely identified. Recently, integer programming has also been used to construct sessions and conduct additional analyses on sessions [25, 26].

Click stream analysis has contributed to new browser technology such as link prefetching [56, 54] from the Mozilla Firefox browser [21] together with enhanced web page caching from the Opera browser [4]. Link prefetching corresponds to the loading of links in the background that can be visited in the future allowing a faster browsing experience. In this case, the access register in the log corresponds to a machine operation which does not reflect human behavior. This increases the difficulty of constructing sessions from web logs.

Sessionization based on web logs passes several processing phases [23]. Firstly, data acquisition is performed during a web server's operation, where for each HTTP request a register is recorded in the web log. During data collection, files are selected and scanned and information stored in temporal repositories for further transformation. Data cleaning consists of selecting only the valid registers: this includes discarding robots, exploits and worm attacks; and finally only html files are selected (discarding multimedia, images, etc). The chunking process [25] splits the large set of valid log registers into smaller sections based on the same IP, agent combination and a time threshold between consecutive registers. The last partition does not ensure unique sessions since multiple web users can share the same IP and agent, but it simplifies further processing. A chunk is the main data unit through which a sessionization algorithm retrieves the web user's trails.

2.5.1   Representation of the Trails

Sessions have different representations [29] depending on what further data mining is to be conducted. At least three historical representations of a session have been used.

Weight usage per pages
Following earlier work on web user profiling [72], sessions are processed to obtain a weight vector $V = [w_1, ..., w_n]$ of normalized time of visits for each page. The vector's dimension is n corresponding to the number of pages in the web site, and the index relates to the corresponding page. Vector entries are zero when the page is not visited and proportional to the time spent on the page when the user visited it. This representation is useful for web user's profiling based on clustering. The weight can be tested with many other representative measures. A binary weighting scheme could be used ($w_i \in \{0,1\}$) that is suitable for association rule analysis [71]. In this case, the weight takes value one if the page is visited during the session. Another variation makes use of prior knowledge of each page [72].

Graph of sessions
[39] Considers the time spent on each page for similarity calculations. The representation of a session is given by the sequence of visited pages. A weighted graph is constructed using a similarity measure and an algorithm of sequence alignment [39]. This representation is further processed using a graph clustering algorithm for web user classification.

Considering text weighting
Other works [99, 48] take into account the semantics of text and the time used. An important page vector is defined containing the page (text content) and the percentage of time spent on each page sorted by time usage and selecting only a fixed number of pages with maximum time usage. This representation is generalized for multimedia content using web objects. The clustering of important page vectors are perform via the SOFM (Self Organized Feature Map) [57] algorithm obtaining categorizations for web user.

2.5.2   Proactive Sessionization

Proactive sessionization directly records a web user's trails but there are some implementation subtilties. A variety of proactive session retrieval methods are described.

2.5.2.1   Cookie Based Sessionization Method

Cookies are a data repository in the web browser that can be read and updated by an embedded program on the web page. This mechanism, usually used for persistence purposes, can uniquely identify the web user by storing a unique identifier and sending this information to the web server (Figure 2.3).

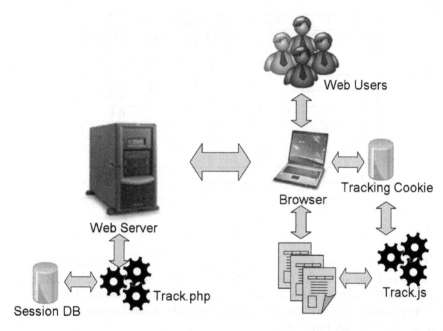

Fig. 2.3 Cookie Sessionization: An embedded script (Track.js) updates the cookie with a unique identification per each user and sends it along with the current URL to the web server web application (Track.php) for recording the page on the session's database

   The method of session processing with cookies is simple in appearance but has it own issues. Security and privacy issues (already mentioned) cause web users to often disable the cookie facilities of their web browsers. Two DOM methods are available for cookie processing, the "onload" method that can be replaced by direct execution of code on HTML and the "onbeforeunload" that is more restricted for browsers executed at the moment of leaving the page. Entering and leaving a web page could be recorded on the web server in order for it to calculate the sequence of pages and time of visit foreach page. A range of browser execution policies provide different pattern access to the same session. Heuristic methods should be used to obtain sessions for some undetermined cases (Table 2.2). For example, if it was registered as an enter event on a page and then a leaving event from another page, the time of visit for both page could be approximated as half of the whole period. A more precise heuristic could estimate the visit time based on similar known sessions.

A futher refining method could be the loading time of the web page. There are 11 combinations of events from the previous page to the the next page; impression occurs in some cases when the page corresponds to the first or last page.

Table 2.2 Simple heuristic for visit time estimation with cookie (1: the event is registered, 0: if not)

Page 1		Page 2		Page 3		
(E)IN	(F)OUT	(A)IN	(B)OUT	(C)IN	(D)OUT	TIME Page 2
1	1	1	1	1	1	B-A
1	1	1	0	1	1	C-A
0	1	1	1	0	1	B-A
1	0	1	1	1	0	B-A
1	1	0	1	1	1	B-F
0	0	1	1	0	0	(D-A)/2
1	0	1	0	1	0	C-A
0	1	0	1	0	1	B-F
1	0	0	1	1	0	(B-E)/2
0	0	1	0	0	0	Indeterminate
0	0	0	1	0	0	Indeterminate

Despite the problem of security and privacy of cookies some advances have been made in order to secure its usage [109]. This consists of automatic validation of cookies for the safety of the web user, based on automatic classification algorithms. Others protocols like P3P [66, 85] relate to the web site declaration of a cookie's degree of private data retrieval.

2.5.2.2 Tracking Application

A spyware application monitors and stores events on the host machine. They can hold links to criminal activity because they can be used to retrieve personal credit card numbers and passwords. These extreme tracking applications are a world wide internet security problem, but some applications have been designed for scientific purposes, for example, AppMonitor [2] that tracks low level events such as mouse clicks on Windows OS for Word and Adobe Reader.

2.5.2.3 Logged Users

Web applications with login authentication can maintain a register of a user's trails on a proprietary database. This can be implemented by storing each page visit during the web user's visit. This is the simplest and most reliable way to track sessions but requires client permission and authentication.

2.5.3   Reactive Sessionization

Sessions obtained from log files are anonymous because web user identity does not appears explicitly in the registers. Of course, this complicates the identification of a web user's trail (Reactive Sessionization). Web users with a similar profile could be accessing the same part of a site at the same time resulting in registers that appears shuffled on Logs. Additionally, web page caching (e.g., Back Button) from browsers and proxy servers can contribute to the cause of log registers going missing. Untangling sessions from Log files requires some assumptions like the maximum time spent by web users on a session, the web site's topology compliance and the semantic content of web pages.

2.5.3.1   Traditional Heuristic Sessionization

Web Logs maintain four important fields for each register: The web user's IP address, the date and time of the request, the url requested before of the current (when activated), and the browser agent identification (Agent) [110]. Using this data a number of processing methods for session retrieval have been used previously. Traditional heuristics start from a partition of the Log register set by grouping according to the same IP number and Agent description. An IP number and agent can not uniquely describe a web user since Internet Service Provider (ISP) multiplex the same IP number over several users employing the network address protocol (NAT).

Time Oriented Heuristic
One of the most popular methods for session retrieval is based on a maximum time a user stays on a web site. After partitioning by IP and Agent, the register is further sliced considering the accumulated time of the visit. There is a tradition of using 30 minutes for the time window [94]. This time is based on empirical experience of the web data mining community. Nowadays, this number does not carry an explanation, but it does seem to provide reasonable results [48]. Another less used approach is to limit the time spent per page.

Web Site Topology Oriented Heuristics
The site topology motivates another heuristic where web users strictly follow the hyperlink structure of a web site. If a register can not be reached from the last register in a web log, it begings a new session [20, 94]. The heuristic scans registers with the same IP and Agent, starting a new session each time a register can not be followed by the previous (Figure 2.5). Of course, this does not uniquely identify the individual path and such a heuristic encounters difficulty when two or more users follow the same path more or less at the same time [25]. This is the case for web sites that have frequently accessed content with only a fewer ways of accessing it. For example, the financial news from a news web site. When a browser or proxy cache is activated, "path competition" can be used to reconstruct the missing registers and conforming a session [94] by selecting the shortest path for the missing registers. If

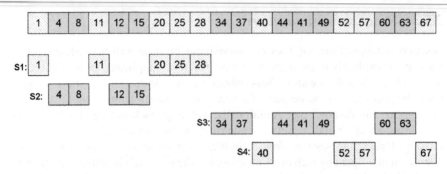

Fig. 2.4 Time Based Sessionization example: a log register indexed by time (second) is segmented in two groups (IP/Agent) generating four session. A timeout occurs after registers 15, 28 and 63

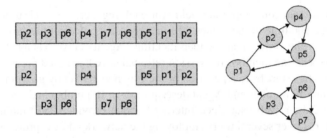

Fig. 2.5 Topology Based Sessionization example: Given the sequence of pages in the log register (left hand) and the web page's structure (right hand) if a page (p3) does not follows the web site hyperlink structure then it start a new session (p4,p7,p5)

the referred field of the log file is activated, the path competition heuristic can be enhanced because the previous pages are provided [65].

2.5.3.2 Ontology Based Sessionization

This method consists of enriching web registers with the semantic information of the visited URL. It is based on the assumption that each web user has a defined purpose that is reflected in the sequence of pages which one visited. The ontology description of each URL must be stated before [52, 51] and a semantic distance matrix Δ_{ij} is calculated from the URL i to j. Sessions are constructed by including the URL that has the nearest semantic distance. Another approach consists of defining predefined sub-trails [55] categorized by ontology.

2.5.3.3 Integer Programming Models

Integer programming is a well known technique used in engineering and operation research for modeling problems that have a combinatorial nature. Substantial progress has been archieved in this particular area in the last 10 years at improving solution time. In the past problemas that took months to resolve can now be solved in minutes [9]. The sessionization problem is inherently combinatoria. Two optimization models for sessionization are presented in this section [25]. The optimization models group log registers from the same IP address and agent. Additional restrictions ensure the link structure of the site is followed in any constructed session. Unlike the traditional heuristics that construct the sessions one at a time, optimization models constructs all sessions simultaneously. Each constructed session from a web server log is an ordered list of log registers where each register can only be used once in only one session. In the same session, register $r1$ can be an immediate predecessor of $r2$ if: the two registers share the same IP address and agent; a link exists from the page requested by $r1$ to the page requested by $r2$; and the request time for register $r2$ is within an allowable time window.

Maximizing a total reward per session
The Sessionization Integer Program (SIP) considers a binary set of variables $\{X_{ros}\}$ that have value one if register r is part of session s in the position o and zero otherwise. If maximizing the number of sessions is considered then the solution is equal to the number of log registers. In this case, a session with only one register satisfies all the previous restriction. Earlier studies on integer programming sessionization [25, 26] shows that considering a linear combination of variables as an objective function for maximization provides results in accordance within the Zipf law for session size (Section 2.2).

$$Maximize \quad \Sigma_{ros} C_{ro} X_{ros}$$

$$Subject \ to:$$

$$\sum_{os} X_{ros} \quad = 1 \qquad \forall r \tag{2.4}$$

$$\sum_{r} X_{ros} \quad \leqslant 1 \qquad \forall o,s \tag{2.5}$$

$$X_{r,o+1,s} \leqslant \Sigma_{r' \in bpage} X_{r'os} \quad \forall r,o,s \tag{2.6}$$

$$X_{ros} \in \{0,1\} \qquad\qquad \forall r,o,s$$
$$X_{ros} = 0 \qquad\qquad \forall s, \ r \in first, \ o > 1$$

The concise formulation of the optimization problem is given in the previous set of equations with a set of restriction from 2.4 to 2.6. The set $bpage_r$ correspond to the set of register r' that can be immediately before the register r in the same session. This is based on the hyperlink precedence of pages, the IP address matching between r and r', the agent matching and time precedence. The set $first$ contains register labels that can be first in a session, and can easily be formed by the condition $r \in first \ if \ bpage_r = \phi$.

The objective function can be understood as a total reward of $\sum_{r,o} {}_L C_{ro}$ per session of size L. The constant C_{ro} should be a monotonic function in o and should reward larger session to avoid the attraction of singleton sessions.

Minimizing the number of sessions
Sessions can be represented as a network flow problem in a directed graph of nodes representing web log registers. The authors in [27] construct a bipartite cardinality matching where (in polynomial time) a solution corresponds to the minimum number of possible sessions can be found for a given web log.

2.5.3.4   Session Analysis

Using integer programming for extracting web user sessions from a web log has the advantage of being a flexible framework for encountering variations of additional information regarading some sessions. Some example follow.

The maximum number of copies of a given session
Finding the maximum number of copies of a given session enable the exploration of any fixed sequence of visited pages. Those patterns are interesting for the study of user habits on a web site. Specific patterns can be tested and ranked by the maximum number of possible sessions for a given web log. Higher ranked patterns should be considered as the most likely sequence of pages from a web log. The formulation correspond to a network flow problem with side constraints [27]. This can be solved quickly, allowing solution of a large number of different patterns within a reasonable time.

The maximum number of sessions of a given size
By adjusting $C_{ro} = 0 \ \forall \ r,o \neq l$ and $C_{ro} = 1 \ \forall \ r,o = l$, one can find the maximum number of sessions of a given size l. This can be used as a measure for the capacity of a web site to capture the attention of the web user.

The maximum number of sessions that pass by a given page in a given order
Some pages on the web site are considered important for navigation and should be present in a large number of sessions. The maximal reward per session can be modified by adjusting $C_{ro} = 1$ when the register r relates to the page and o is the order, and $C_{ro} = 0$ otherwise.

Considering the effect of cache devices on web logs
As presented in section 2.2 the use of the back button in web browsing produces missing registers in web logs. The maximal reward per session can be modified by introducing a new variable Y_{ros} corresponding to the use of a back button action [26]. A session in this case could be identified to include a session with back button usage and constraints can be added on the total number (or percent) of sessions which includes the use of the back button.

2.5.4 Sessions in Dynamic Environments

Static assumptions for web usage mining are a common hypothesis for data pre-processing. However in web application like CRM and recommender systems dynamic content is not the exception but the rule. In this dynamic context, the notion of dynamic URL becomes a valid representation [76]. Query parameter values from the URL and application database contribute to map the content with the dynamic URL, and convert it into a hierarchical semantically structure $(i, j) | i : parent, j : child$ of semantic label.

The evolution of a web site produces changes in the behavior of the web users. Thus, web user sessions should be comparable to a period of time where changes can be considered minimal. Those periods $\{T_1, ..., T_k\}$ depend on web site managers' updating procedures which need to be defined. Then, a given period of sessionization has to be performed using the available semantic labels in the page. Periods of sessionization can then be compared to the analysis of the user profile evolution [76]. The use of semantic labelling provides a comparison between the sessions belonging to different version of the same web site.

2.5.5 Identifying Session Outliers

An important phase of data pre-processing is data cleaning. A first stage of the sessionization process cleans the Log file erasing register from robot, viruses/worms, hacking attempt and others. But in general not all of this unwanted data can be removed. The sessions set could be refined detecting different modes of uncommon behavior [88, 52]. Some recent studies [88] use the 1% tail of the Malanobis distance to locate sessions indicating unusual modes of behaviors.Others studies relate to semantic characteristics of the sessions [52] which are obtained using an Ontology-Oriented heuristic.

2.6 Summary

Web data is a complex and noisy data source, yet empirical regularity rules its statistical description. Several pre-processing techniques have been developed in the last ten years to support web mining and address the changing characteristics of the Web. The primary web data extracted are the hyperlink structure, the content, and the usage of a web site. All have some collection issues. A web crawler collects the hyperlink structure but the time changing characteristic of the Web must be handled with page selection methods, revisit schedules, a politeness policy, and parallel processing. The challenges with web content are determining a weighting scheme as well as dealing with the visual representation and the dynamism of a website. Web usage data can be obtained indirectly from web logs or by direct retrieval. Integer programming has recently shown promise as an indirect method. Data preparation is a significant effort and a necessary cornerstone for web mining.

Acknowledgements. This work was supported partially by the FONDEF project DO8I-1015 and the *National Doctoral Grant* from *Conicyt Chile*. We are grateful to Luis E. Dujovne for very fruitful discussion on Web Site Key Objects.

References

1. Adar, E., Teevan, J., Dumais, S., Elsas, J.: The web changes everything: understanding the dynamics of web content. In: WSDM 2009: Proceedings of the Second ACM International Conference on Web Search and Data Mining, pp. 282–291. ACM Press, New York (2009)
2. Alexander, J.: Understanding and improving navigation within electronic documents. Ph.D. thesis, University of Canterbury, Christchurch, New Zealand (2009)
3. Alexander, J., Cockburn, A.: An empirical characterisation of electronic document navigation. In: GI 2008: Proceedings of graphics interface 2008, pp. 123–130. Canadian Information Processing Society, Toronto (2008)
4. ASA, O.S.: Opera browser, http://www.opera.com
5. Baeza-Yates, R., Castillo, C., Efthimiadis, E.: Characterization of national web domains. ACM Transactions on Internet Technology 7(2) (2007)
6. Baeza-Yates, R., Poblete, B.: Dynamics of the chilean web structure. Comput. Netw. 50(10), 1464–1473 (2006)
7. Bayir, M., Toroslu, I., Cosar, A., Fidan, G.: Smart miner: a new framework for mining large scale web usage data. In: WWW 2009: Proceedings of the 18th international conference on World wide web, pp. 161–170. ACM Press, New York (2009)
8. Bhamidipati, N.L., Pal, S.K.: Stemming via distribution-based word segregation for classification and retrieval. IEEE Transactions on Systems, Man, and Cybernetics, Part B 37(2), 350–360 (2007)
9. Bixby, R.E.: Solving real-world linear programs: A decade and more of progress. Operations Research 50(1), 3–15 (2002)
10. Burget, R., Rudolfova, I.: Web page element classification based on visual features. In: Asian Conference on Intelligent Information and Database Systems, vol. 0, pp. 67–72 (2009)
11. Castells, P., Fernandez, M., Vallet, D.: An adaptation of the vector-space model for ontology-based information retrieval. IEEE Trans. on Knowl. and Data Eng. 19(2), 261–272 (2007)
12. Castillo, C.: Effective web crawling. Ph.D. thesis, University of Chile, Santiago, Chile (2004)
13. Catledge, L.D., Pitkow, J.E.: Characterizing browsing strategies in the world-wide web. In: Computer Networks and ISDN Systems, pp. 1065–1073 (1995)
14. Chakrabarti, D., Faloutsos, C.: Graph mining: Laws, generators, and algorithms. ACM Comput. Surv. 38(1), 2 (2006)
15. Chakrabarti, D., Kumar, R., Punera, K.: Page-level template detection via isotonic smoothing. In: WWW 2007: Proceedings of the 16th international conference on World Wide Web, pp. 61–70. ACM Press, New York (2007)
16. Chakrabarti, S., Dom, B.E., Kumar, S.R., Raghavan, P., Rajagopalan, S., Tomkins, A., Gibson, D., Kleinberg, J.: Mining the web's link structure. Computer 32(8), 60–67 (1999)

17. Chang, F., Dean, J., Ghemawat, S., Hsieh, W.C., Wallach, D.A., Burrows, M., Chandra, T., Fikes, A., Gruber, R.E.: Bigtable: a distributed storage system for structured data. In: OSDI 2006: Proceedings of the 7th USENIX Symposium on Operating Systems Design and Implementation, p. 15. USENIX Association, Berkeley (2006)
18. Cho, J., Garcia-Molina, H.: The evolution of the web and implications for an incremental crawler. In: VLDB 2000: Proceedings of the 26th International Conference on Very Large Data Bases, pp. 200–209. Morgan Kaufmann Publishers Inc., San Francisco (2000)
19. Cho, J., Garcia-Molina, H.: Estimating frequency of change. ACM Trans. Internet Technol. 3(3), 256–290 (2003)
20. Cooley, R., Mobasher, B., Srivastava, J.: Data preparation for mining world wide web browsing patterns. Knowledge and Information Systems 1, 5–32 (1999)
21. Corporation, M.: Mozilla firefox browser, http://www.mozilla.org
22. Coull, S.E., Collins, M.P., Wright, C.V., Monrose, F., Reiter, M.K.: On web browsing privacy in anonymized netflows. In: SS 2007: Proceedings of 16th USENIX Security Symposium on USENIX Security Symposium, pp. 1–14. USENIX Association, Berkeley (2007)
23. Das, R., Turkoglu, I.: Creating meaningful data from web logs for improving the impressiveness of a website by using path analysis method. Expert Syst. Appl. 36(3), 6635–6644 (2009)
24. Debnath, S., Mitra, P., Pal, N., Giles, C.L.: Automatic identification of informative sections of web pages. IEEE Trans. on Knowl. and Data Eng. 17(9), 1233–1246 (2005)
25. Dell, R.F., Román, P.E., Velásquez, J.D.: Web user session reconstruction using integer programming. In: Procs. of The 2008 IEEE/WIC/ACM International Conference on Web Intelligence, Sydney, Australia, pp. 385–388 (2008)
26. Dell, R.F., Román, P.E., Velásquez, J.D.: User session reconstruction with back button browsing. In: Velásquez, J.D., Ríos, S.A., Howlett, R.J., Jain, L.C. (eds.) Knowledge-Based and Intelligent Information and Engineering Systems. LNCS, vol. 5711, pp. 326–332. Springer, Heidelberg (2009)
27. Dell, R.F., Román, P.E., Velásquez, J.D.: Optimization models for construction of web user sessions. Working Paper (2010)
28. Demartini, G., Firan, C.S., Iofciu, T., Nejdl, W.: Semantically enhanced entity ranking. In: Bailey, J., Maier, D., Schewe, K.-D., Thalheim, B., Wang, X.S. (eds.) WISE 2008. LNCS, vol. 5175, pp. 176–188. Springer, Heidelberg (2008)
29. Demir, G.N., Goksedef, M., Etaner-Uyar, A.S.: Effects of session representation models on the performance of web recommender systems. In: ICDEW 2007: Proceedings of the 2007 IEEE 23rd International Conference on Data Engineering Workshop, pp. 931–936. IEEE Computer Society Press, Washington (2007)
30. Desikan, P., Srivastava, J.: Mining temporally evolving graphs. In: Mobasher, B., Nasraoui, O., Liu, B., Masand, B. (eds.) WebKDD 2004. LNCS (LNAI), vol. 3932, pp. 1–17. Springer, Heidelberg (2004)
31. Dill, S., Kumar, R., Mccurley, K., Rajagopalan, S., Sivakumar, D., Tomkins, A.: Self-similarity in the web. ACM Trans. Internet Technol. 2(3), 205–223 (2002)
32. Dujovne, L.E., Velásquez, J.D.: Design and implementation of a methodology for identifying website keyobjects. In: Velásquez, J.D., Ríos, S.A., Howlett, R.J., Jain, L.C. (eds.) Knowledge-Based and Intelligent Information and Engineering Systems. LNCS, vol. 5711, pp. 301–308. Springer, Heidelberg (2009)
33. Eguchi, S., Copas, J.: Interpreting kullback-leibler divergence with the neyman-pearson lemma. J. Multivar. Anal. 97(9), 2034–2040 (2006)

34. Fetterly, D., Manasse, M., Najork, M., Wiener, J.: A large-scale study of the evolution of web pages. In: WWW 2003: Proceedings of the 12th international conference on World Wide Web, pp. 669–678. ACM Press, New York (2003)
35. Gaugaz, J., Zakrzewski, J., Demartini, G., Nejdl, W.: How to trace and revise identities. In: Aroyo, L., Traverso, P., Ciravegna, F., Cimiano, P., Heath, T., Hyvönen, E., Mizoguchi, R., Oren, E., Sabou, M., Simperl, E. (eds.) ESWC 2009. LNCS, vol. 5554, pp. 414–428. Springer, Heidelberg (2009)
36. Ghani, R., Jones, R., Mladenic, D.: Mining the web to create minority language corpora. In: CIKM 2001: Proceedings of the tenth international conference on Information and knowledge management, pp. 279–286. ACM Press, New York (2001)
37. Görnitz, N., Kloft, M., Brefeld, U.: Active and semi-supervised data domain description. In: ECML PKDD 2009: Proceedings of the European Conference on Machine Learning and Knowledge Discovery in Databases, pp. 407–422. Springer, Heidelberg (2009)
38. Granka, L., Feusner, M., Lorigo, L.: Eye monitoring in online search. In: Hammoud, R., Ohno, T. (eds.) Passive Eye Monitoring, Signals and Communication Technology, Part VI, pp. 347–372. Springer, Heidelberg (2008)
39. Gündüz, C., Özsu, M.T.: A web page prediction model based on click-stream tree representation of user behavior. In: KDD 2003: Proceedings of the ninth ACM SIGKDD international conference on Knowledge discovery and data mining, pp. 535–540. ACM Press, New York (2003)
40. Hand, D.: Statistics and data mining: intersecting disciplines. SIGKDD Explor. Newsl. 1(1), 16–19 (1999)
41. Hensman, S.: Construction of conceptual graph representation of texts. In: HLT-NAACL 2004: Proceedings of the Student Research Workshop at HLT-NAACL 2004, vol. XX, pp. 49–54. Association for Computational Linguistics, Morristown (2004)
42. Huberman, B., Pirolli, P., Pitkow, J., Lukose, R.M.: Strong regularities in world wide web surfing. Science 280(5360), 95–97 (1998)
43. Huberman, B., Wu, F.: The economics of attention: maximizing user value in information-rich environments. In: ADKDD 2007: Proceedings of the 1st international workshop on Data mining and audience intelligence for advertising, pp. 16–20. ACM Press, New York (2007)
44. Iachello, G., Hong, J.: End-user privacy in human-computer interaction. Found. Trends Hum.-Comput. Interact. 1(1), 1–137 (2007)
45. Ipeirotis, P., Gravano, L.: When one sample is not enough: improving text database selection using shrinkage. In: SIGMOD 2004: Proceedings of the 2004 ACM SIGMOD international conference on Management of data, pp. 767–778. ACM Press, New York (2004)
46. Janzen, S., Maass, W.: Ontology-based natural language processing for in-store shopping situations. In: ICSC 2009: Proceedings of the 2009 IEEE International Conference on Semantic Computing, pp. 361–366. IEEE Computer Society, Washington (2009)
47. Jatowt, A., Ishizuka, M.: Temporal multi-page summarization. Web Intelli. and Agent Sys. 4(2), 163–180 (2006)
48. Velásquez, J.D., Palade, V.: Adaptive web sites: A knowledge extraction from web data approach. IOS Press, Amsterdam (2008)
49. Jin, W., Srihari, R.K.: Graph-based text representation and knowledge discovery. In: SAC 2007: Proceedings of the 2007 ACM symposium on Applied computing, pp. 807–811. ACM, New York (2007)

50. Joachims, T., Granka, L., Pan, B., Hembrooke, H., Radlinski, F., Gay, G.: Evaluating the accuracy of implicit feedback from clicks and query reformulations in web search. ACM Trans. Inf. Syst. 25(2), 7 (2007)
51. Jung, J.J.: Ontology-based partitioning of data steam for web mining: A case study of web logs. In: ICCS 2004, 4th International Conference, Proceedings, Part I, June 6-9, 2004, Kraków, Poland, pp. 247–254 (2004)
52. Jung, J.J., Jo, G.S.: Semantic outlier analysis for sessionizing web logs. In: ECML/P-KDD Conference, pp. 13–25 (2004)
53. Ke, Y., Deng, L., Ng, W., Lee, D.: Web dynamics and their ramifications for the development of web search engines. Comput. Netw. 50(10), 1430–1447 (2006)
54. Khan, J.I., Tao, Q.: Exploiting webspace organization for accelerating web prefetching. Web Intelli. and Agent Sys. 3(2), 117–129 (2005)
55. Khasawneh, N., Chan, C.: Active user-based and ontology-based web log data preprocessing for web usage mining. In: 2006 IEEE / WIC / ACM International Conference on Web Intelligence (WI 2006), Hong Kong, China, pp. 325–328. IEEE Computer Society, Los Alamitos (2006)
56. Kim, Y., Kim, J.: Web prefetching using display-based prediction. In: WI 2003: Proceedings of the 2003 IEEE/WIC International Conference on Web Intelligence, p. 486. IEEE Computer Society, Washington (2003)
57. Kohonen, T.: Self-organized formation of topologically correct feature maps, pp. 509–521 (1988)
58. Kryssanov, V., Kakusho, K., Kuleshov, E., Minoh, M.: Modeling hypermedia-based communication. Information Sciences 174(1-2), 37–53 (2005)
59. Lan, M., Tan, C.L., Low, H.B., Sung, S.Y.: A comprehensive comparative study on term weighting schemes for text categorization with support vector machines. In: WWW 2005: Special interest tracks and posters of the 14th international conference on World Wide Web, pp. 1032–1033. ACM Press, New York (2005), http://doi.acm.org/10.1145/1062745.1062854
60. Lan, M., Tan, C.L., Su, J., Lu, Y.: Supervised and traditional term weighting methods for automatic text categorization. IEEE Trans. Pattern Anal. Mach. Intell. 31(4), 721–735 (2009)
61. Langford, D.: Internet ethics. MacMillan Press Ltd., Basingstoke (2000)
62. Lansey, J.C., Bukiet, B.: Internet search result probabilities, heaps' law and word associativity. Journal of Quantitative Linguistics 16(1), 40–66 (2005)
63. Leijenhorst, D.V., der Weide, T.V.: A formal derivation of heaps' law. Inf. Sci. Inf. Comput. Sci. 170(2-4), 263–272 (2005)
64. Levene, M., Borges, J., Loizou, G.: Zipf's law for web surfers. Knowl. Inf. Syst. 3(1), 120–129 (2001)
65. Li, Y., Feng, B., Mao, Q.: Research on path completion technique in web usage mining. In: International Symposium on Computer Science and Computational Technology, vol. 1, pp. 554–559 (2008)
66. Linn, J.: Technology and web user data privacy: A survey of risks and countermeasures. IEEE Security and Privacy 3(1), 52–58 (2005)
67. Liu, B.: Web Data Mining: Exploring Hyperlinks, Contents, and Usage Data (Data-Centric Systems and Applications), 1st edn. (2007); corr. 2nd printing edn. Springer, Heidelberg (2009)
68. Manning, C.D., Schutze, H.: Fundation of Statistical Natural Language Processing. MIT Press, Cambridge (1999)
69. Maynor, D.: Metasploit Toolkit for Penetration Testing, Exploit Development, and Vulnerability Research, 1st edn. Syngress (2007)

70. Mobasher, B.: Web usage mining. In: Liu, B. (ed.) Web Data Mining: Exploring Hyperlinks, Contents and Usage Data, ch. 12. Springer, Heidelberg (2006)

71. Mobasher, B., Dai, H., Luo, T., Nakagawa, M.: Effective personalization based on association rule discovery from web usage data. In: WIDM 2001: Proceedings of the 3rd international workshop on Web information and data management, pp. 9–15. ACM Press, New York (2001)

72. Mobasher, B., Dai, H., Luo, T., Nakagawa, M.: Discovery and evaluation of aggregate usage profiles for web personalization. Data Min. Knowl. Discov. 6(1), 61–82 (2002)

73. Moloney, M., Bannister, F.: A privacy control theory for online environments. In: HICSS 2009: Proceedings of the 42nd Hawaii International Conference on System Sciences, pp. 1–10. IEEE Computer Society, Washington (2009)

74. Mori, T.: Information gain ratio as term weight: the case of summarization of ir results. In: Proceedings of the 19th international conference on Computational linguistics, pp. 1–7. Association for Computational Linguistics, Morristown, NJ, USA (2002)

75. Nadeax, D.: Semi-supervised named entity recognition: Learning to recognize 100 entity types with little supervision. Ph.D. thesis, University of Ottawa, Ottawa, Canada (2007)

76. Nasraoui, O., Soliman, M., Saka, E., Badia, A., Germain, R.: A web usage mining framework for mining evolving user profiles in dynamic web sites. IEEE Trans. on Knowl. and Data Eng. 20(2), 202–215 (2008)

77. Obendorf, H., Weinreich, H., Herder, E., Mayer, M.: Web page revisitation revisited: implications of a long-term click-stream study of browser usage. In: CHI 2007: Proceedings of the SIGCHI conference on Human factors in computing systems, pp. 597–606 (2007)

78. Olston, C., Pandey, S.: Recrawl scheduling based on information longevity. In: WWW 2008: Proceeding of the 17th international conference on World Wide Web, pp. 437–446. ACM Press, New York (2008)

79. Pal, S.K., Talwar, V., Mitra, P.: Web mining in soft computing framework: Relevance, state of the art and future directions. IEEE Transactions on Neural Networks 13, 1163–1177 (2002)

80. Peña-Ortiz, R., Sahuquillo, J., Pont, A., Gil, J.: Dweb model: Representing web 2.0 dynamism. Comput. Commun. 32(6), 1118–1128 (2009)

81. Popov, B., Kiryakov, A., Ognyanoff, D., Manov, D., Kirilov, A.: Kim – a semantic platform for information extraction and retrieval. Nat. Lang. Eng. 10(3-4), 375–392 (2004)

82. Porter, M.F.: An algorithm for suffix stripping. Electronic Library and Electronic Systems 40, 211–218 (2006)

83. Qi, X., Davison, B.: Web page classification: Features and algorithms. ACM Comput. Surv. 41(2), 1–31 (2009)

84. Radlinski, F., Kurup, M., Joachims, T.: How does clickthrough data reflect retrieval quality? In: CIKM 2008: Proceeding of the 17th ACM conference on Information and knowledge management, pp. 43–52. ACM Press, New York (2008)

85. Reay, I.K., Beatty, P., Dick, S., Miller, J.: A survey and analysis of the p3p protocol's agents, adoption, maintenance, and future. IEEE Transactions on Dependable and Secure Computing 4, 151–164 (2007)

86. Román, P.E., Velásquez, J.D.: Dynamic stochastic model applied to the analysis of the web user behavior. In: 6th Atlantic Web Intelligence Conference, AWIC 2009, Prague, CZECH Republic, pp. 31–40 (2009)

87. Rugaber, S., Harel, N., Govindharaj, S., Jerding, D.: Problems modeling web sites and user behavior. In: WSE 2006: Proceedings of the Eighth IEEE International Symposium on Web Site Evolution, pp. 83–94. IEEE Computer Society Press, Washington (2006)

88. Sadagopan, N., Li, J.: Characterizing typical and atypical user sessions in clickstreams. In: WWW 2008: Proceeding of the 17th international conference on World Wide Web, pp. 885–894. ACM Press, New York (2008)

89. Sebastiani, F.: Machine learning in automated text categorization. ACM Comput. Surv. 34(1), 1–47 (2002)

90. Shehata, S.: A wordnet-based semantic model for enhancing text clustering. In: ICDMW 2009: Proceedings of the 2009 IEEE International Conference on Data Mining Workshops, pp. 477–482. IEEE Computer Society, Washington (2009)

91. Snásel, V., Kudelka, M.: Web content mining focused on named objects. In (IHCI) First International Conference on Intelligent Human Computer Interaction, pp. 37–58. Springer, India (2009)

92. Soares, M.V.B., Prati, R.C., Monard, M.C.: Improvement on the porter's stemming algorithm for portuguese. IEEE Latin America Transaction 7(4), 472–477 (2009)

93. Spaniol, M., Denev, D., Mazeika, A., Weikum, G., Senellart, P.: Data quality in web archiving. In: WICOW 2009: Proceedings of the 3rd workshop on Information credibility on the web, pp. 19–26. ACM Press, New York (2009)

94. Spiliopoulou, M., Mobasher, B., Berendt, B., Nakagawa, M.: A framework for the evaluation of session reconstruction heuristics in web-usage analysis. Informs Journal on Computing 15(2), 171–190 (2003)

95. Tauscher, L., Greenberg, S.: Revisitation patterns in world wide web navigation. In: Procs. of the Conference on Human Factors in Computing Systems, Atlanta, USA, pp. 22–27 (1997)

96. Tsatsaronis, G., Varlamis, I., Nørvg, K.: An experimental study on unsupervised graph-based word sense disambiguation. In: Gelbukh, A. (ed.) Computational Linguistics and Intelligent Text Processing. LNCS, vol. 6008, pp. 184–198. Springer, Heidelberg (2010)

97. Ullrich, C., Borau, K., Luo, H., Tan, X., Shen, L., Shen, R.: Why web 2.0 is good for learning and for research: principles and prototypes. In: WWW 2008: Proceeding of the 17th international conference on World Wide Web, pp. 705–714. ACM Press, New York (2008)

98. Urbansky, D., Feldmann, M., Thom, J.A., Schill, A.: Entity extraction from the web with webknox. In: 6th Atlantic Web Intelligence Conference (AWIC), Prague, Czech Republic (2009)

99. Velásquez, J.D., Yasuda, H., Aoki, T., Weber, R., Vera, E.: Using self organizing feature maps to acquire knowledge about visitor behavior in a web site. In: Palade, V., Howlett, R.J., Jain, L. (eds.) KES 2003. LNCS, vol. 2773, pp. 951–958. Springer, Heidelberg (2003)

100. Wang, J., Wu, X., Zhang, C.: Support vector machines based on kmeans clustering for real time business intelligence systems. Int. J. Bus. Intell. Data Min. 1(1), 54–64 (2005)

101. Wang, Y., Hodges, J.: Document clustering with semantic analysis. In: HICSS 2006: Proceedings of the 39th Annual Hawaii International Conference on System Sciences, p. 54.3. IEEE Computer Society, Washington (2006)

102. Weinreich, H., Obendorf, H., Herder, E., Mayer, M.: Off the beaten tracks: exploring three aspects of web navigation. In: WWW 2006: Proceedings of the 15th international conference on World Wide Web, pp. 133–142. ACM Press, New York (2006)

103. Weinreich, H., Obendorf, H., Herder, E., Mayer, M.: Not quite the average: An empirical study of web use. ACM Trans. Web 2(1), 1–31 (2008)

104. White, R.W.: Investigating behavioral variability in web search. In. Proc. WWW, pp. 21–30 (2007)

105. Wittek, P., Darányi, S., Tan, C.L.: Improving text classification by a sense spectrum approach to term expansion. In: CoNLL 2009: Proceedings of the Thirteenth Conference on Computational Natural Language Learning, pp. 183–191. Association for Computational Linguistics, Morristown (2009)
106. Won, S., Jin, J., Hong, J.: Contextual web history: using visual and contextual cues to improve web browser history. In: CHI 2009: Proceedings of the 27th international conference on Human factors in computing systems, pp. 1457–1466. ACM Press, New York (2009)
107. Yan, X., Zhang, C., Zhang, S.: Toward databases mining: Pre-processing collected data. Applied Artificial Intelligence 17(5-6), 545–561 (2003)
108. Yu, L., Wang, S., Lai, K.: An integrated data preparation scheme for neural network data analysis. IEEE Transactions on Knowledge and Data Engineering 18, 217–230 (2006)
109. Yue, C., Xie, M., Wang, H.: Automatic cookie usage setting with cookiepicker. In: DSN 2007: Proceedings of the 37th Annual IEEE/IFIP International Conference on Dependable Systems and Networks, pp. 460–470. IEEE Computer Society Press, Washington (2007)
110. Zawodny, J.D.: Linux apache web server administration. Sybex, 2 edn. (2002)
111. Zhang, Z., Chen, J., Li, X.: A preprocessing framework and approach for web applications. J. Web Eng. 2(3), 176–192 (2004)

Chapter 3
Web Pattern Extraction and Storage

Víctor L. Rebolledo, Gastón L'Huillier, and Juan D. Velásquez

Abstract. Web data provides information and knowledge to improve the web site content and structure. Indeed, it eventually contains knowledge which suggests changes that makes a web site more efficient and effective to attract and retain visitors. Making use of a Data Webhouse or a web analytics solution, it is possible to store statistical information concerning the behaviour of users in a website. Likewise, through applying web mining algorithms, interesting patterns can be discovered, interpreted and transformed into useful knowledge. On the other hand, web data include quantities of irrelevant but complex data preprocessing that must be applied in order to model and understand visitor browsing behaviour. Nevertheless, there are many ways to pre-process web data and model the browsing behaviour, hence different patterns can be obtained depending on which model is used. In this sense, a knowledge representation is necessary to store and manipulate web patterns. Generally, different patterns are discovered by using distinct web mining techniques on web data with dissimilar treatments. Consequently, patterns meta-data are relevant to manipulate the discovered knowledge. In this chapter, topics like feature selection, web mining techniques, models characterisation and pattern management will be covered in order to build a repository that stores patterns' meta-data. Specifically, a Pattern Webhouse that facilitates knowledge management in the web environment.

Víctor L. Rebolledo · Gastón L'Huillier · Juan D. Velásquez
Department of Industrial Engineering, University of Chile, República 701, Santiago, Chile
e-mail: vireboll@ing.uchile.cl, glhuilli@dii.uchile.cl,
    jvelasqu@dii.uchile.cl

J.D. Velásquez and L.C. Jain (Eds.): Advanced Techniques in Web Intelligence – 1, SCI 311, pp. 49–77.
springerlink.com   © Springer-Verlag Berlin Heidelberg 2010

3.1   Introduction

According to the *Web Intelligence Consortium* (WIC)[1], "*Web Intelligence* (WI) has
been recognised as a new direction for scientific research and development to explore the fundamental roles as well as practical impacts of *Artificial Intelligence*
(AI)[2] and advanced *Information Technology* (IT)[3] on the next generation of Web-
empowered products, systems, services, and activities"

In other words, WI seeks ways to evolve from a Web of data to a Web of knowledge and wisdom. Just as you can see in Figure 3.1, web data source must be understood, particularly, web server and browser interactions. So new ways of pre-
processing web data need to be discovered in order to extract information, which
must be stored in repositories or Data Web-houses [29]. Nevertheless, information
is not enough to make intelligent decisions, therefore knowledge is required. [54].

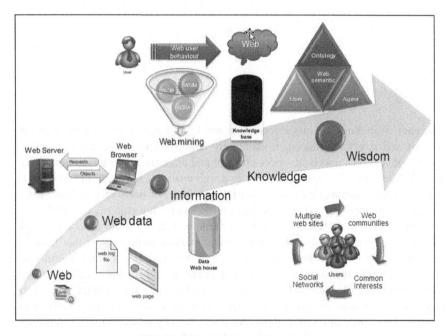

Fig. 3.1 Web Intelligence Overview

---

[1] It is an international, non-profit organization dedicated to advance world-wide scientific research and industrial development in the field of Web Intelligence (WI) `wi-consortium.`
`org`

[2] Knowledge representation, planning, knowledge discovery and data mining, intelligent
agents, and social network intelligence.

[3] Wireless networks, ubiquitous devices, social networks, wisdom Web, and data/knowledge
grids.

Thanks to Web Mining techniques, web user browsing behaviour can be modelled and studied. Indeed, mining web data enables one to find useful and unknown knowledge through patterns which must be correctly interpreted [10]. However, web data need to be pre-processed to apply web mining algorithms whose results depends on the treatment applied to data sources [35]. Moreover, web data is always changing: users change their browsing behaviour and web site content and structure are modified to attract new users. In this sense, the discovered knowledge may become obsolete in a short period of time, therefore Knowledge Base is required to store and manipulate relevant patterns [56].

Finally, knowledge must be managed, for which there are different approaches. One of which is knowledge representation based on ontologies written in XML-based languages. This representation is the basis of **Semantic Web**[4], whose purpose is to enable the Web to resolve requests from people and machines through artificial systems.

Web Intelligence systems will not be limited to extracting information from web data, they can also extract, store and manage knowledge. In fact, they will be able to understand visitors and, consequently, make *"intelligent"* decisions and recommendations based on their interactions.

3.1.1   From Data to Knowledge

Many times, the terms *data*, *information*, and *knowledge* are indistinctly used, due to incorrect interpretations of their meanings. According to Davenport & Prusak in their book *Working Knowledge* [11]:

- **Data** are *primary elements of information* that are irrelevant for decision making. In other words, they can be seen as a discrete group of values that do not mention anything about the nature of the context and furthermore do not provide any guidance. For example, this could be a list of phone numbers.
- **Information** corresponds to a group of processed data that have a meaning, relevance, purpose and context, and therefore, they are *useful for decision making*. In order to obtain information, data must be put into context, categorised, processed, corrected, edited and condensed so that they make sense to decision maker.
- **Knowledge** is composed of a *mixture of experience, values, information and know-how*, which serves as a framework for the incorporation of new experiences and information. The knowledge is originated and applied to the experts mind[5]. In organisations, it is frequently witnessed not only in documents or data warehouses, but also in organisational routines, processes and standards. To obtain knowledge from information, one must compare it with other elements to predict consequences, thus looking for underlying connections or searching for an expert's interpretation.

---

[4] It is an evolving development of the Web in which the meaning (semantics) of information and services is defined.

[5] The experts' minds are source of knowledge and they are able to apply it.

When users are browsing, they visits different pages that contain data in various formats: text, images, multimedia files and so on [56]. For example, if this user looks for information about a certain topic making use of search engines, the results must be put in context [11]. Firstly, web user read the content of pages obtaining information related to the topic [39]. While the user learns more about the topic, he could request new questions, resulting in new enquiries to search engines which provide new pages to users to read. Eventually, they will be able to obtain knowledge about the topic. However, the process could be lengthy and tedious due to the user only obtaining knowledge once they have read every page provided by the search engine.

The previous process will be repeated every time a user wants to obtain information or knowledge through the Web [39]. For some people this process is simpler than for others, however, achieving a high degree of efficiency demands the use of search engines for a certain period of time before becoming an experienced user [57]. In other words, users should adapt to the Web to obtain what they want. In that sense, knowledge contained in the Web must be represented in order to reduce the cost of obtaining it.

3.1.2   About Knowledge Representation

The human mind's mechanism for storing and retrieving knowledge is fairly transparent to us. For example, when an "orange" was memorized it was examined first, thought about for some time and perhaps eaten. During this process, the orange's essential qualities are stored: size, colour, smell, taste, texture, etc. Consequently, when the word "orange" is heard, our senses are activated from within [3].

Unfortunately, computers are not able to build this kind of representation by themselves. Instead of gathering knowledge, computers must rely on human beings to place knowledge directly into their memories. While programming enables us to determine how the computer performs, we must decide on ways to represent information, knowledge, and inference techniques within it.

As field research, *Knowledge representation* (KR) is the study of how knowledge of the world can be represented and what kind of reasoning can be achieved with regards to it. In other words, it is "how an entity sees the world", understands a situation and prepares satisfactory action. In short, it is the ability to create new perceptions from old ones.

According to Davis et al. in their paper "What is Knowledge Representation", the notion of this topic can best be understood in terms of five distinct roles it plays [13, 54]:

- KR is fundamentally a surrogate or substitute for the elements that compose the external world, used to enable an entity to determine consequences of applying an action, i.e., by reasoning about the world rather than taking action.
- KR should answer the question: *In what terms should I think about the world?*. So, KR is a set of ontological commitments, of which the choice of representation must be made. All knowledge representations are approximations of reality.

In this sense, KR is a method, consisting of criteria such as "what we want to perceive" as a filter. For example, a general representation of voice production is to think in terms of a semi-periodic source (glottal excitation), vocal tract, lip radiation and the integration of these components. However, if the focus is on vocal track dynamics only, it can be explained as "the glottal excitation wave hits with the walls of the vocal track, generating a composed wave", which can be understood as a set of ontological commitments.

- KR is a fragmentary theory of intelligent reasoning, expressed in terms of three components:

    - The representation's fundamental conception of intelligent reasoning.
    - The set of inferences to represent sanctions.
    - The set of recommendations implied by the inference.

    For example, in the classic economic theory, consumers make intelligent decisions based on the information about product characteristics and prices. When the price changes, so does the intention to purchase it.
- KR is a medium for pragmatically efficient computation. As thinking machines are in essence computational processes, the challenge to efficient programming is to properly represent the world. Independent of the language used, it is necessary to correctly represent the problem and create a manageable and efficient code, i.e., not requiring redundant and high data processing capacities.
- KR is a medium of human expression. Knowledge is expressed through a medium such as the spoken language, written text, arts, etc. A communication interface is necessary if a human being is to interact with an intelligent system.

In order to provide the five before mentioned roles, KR is a multidisciplinary subject that applies theories and techniques from three other fields:

1. *Logic* provide the formal structure and rules of inference.
2. *Ontology* defines the kind of things that exist in the application domain.
3. *Computation* supports the applications that distinguish knowledge representation from a simple concept.

This combination represents the main initiative to evolve from a Web of data to a Web of knowledge. Indeed, this is the final purpose of **Semantic Web** which is an initiative led by *World Wide Web Consortium* (W3C) that aims to define standards in order to add descriptive formal content to the web pages. This content will be invisible for the users and will enable machines to understand and manipulate the content of each of the pages to provide intelligent services. In this sense, there are diverse trade-offs when we one attempt to represent knowledge. Many times the representational adequacy and fidelity, given by Logic and Ontology, opposed the Computational cost. In dynamic environments with stochastic information these trade-offs are even more pronounced.

On the other hand, knowledge can also be seen as *patterns discovered from data bases by using data mining techniques*. Indeed, patterns can be defined as knowledge artefacts, providing a compact and semantically rich representation of a huge quantity of heterogeneous raw data. However, due to the specific characteristics of patterns (heterogeneous and voluminous), *ad hoc* systems are required for pattern management in order to model, store, retrieve, and manipulate patterns in an effective and efficient way.

In order to make pattern management, different approaches can be founded:

- The design of a *Data Mining Query Language* for mining different kinds of knowledge in relational databases. The idea is to provide an standardised language as SQL to extract patterns so if SQL was successful in extracting information in databases, a new language can resolve the knowledge extraction issue. Examples of this specific language are **DMQL** [23], **MINE RULE** [40], and **MineSQL** (MSQL) [26]
- Commercial initiatives as **PMML**[6], an XML notation represents and describes data mining and statistics models, as well as some of the operations required for cleaning and transforming data prior to modelling. The idea is to provide enough infrastructure for an application to be able to produce a model (the PMML producer) and another one to consume it (the PMML consumer) simply by reading the PMML file [21].
- Build a *Pattern Base Management System* (PBMS) [7]. The idea is formally to define the logical foundations for the global setting of pattern management through a model that covers data, patterns, and their intermediate mappings. Moreover a formalism for pattern specification along with safety restrictions and introduce predicates for comparing patterns and query operators. This system was finally implemented as a prototype called **PSYCHO** [8, 50]

With respecto to a different approach, this chapter focus on characterise web mining models with the purpose of building a repository that facilitates the manipulation of knowledge contained in the discovered patterns. The idea is to store the patterns and meta-data associated with them making use of *Data Warehouse* architecture. Unlike *Data Webhouses*[7] that store statistical information about web users behaviour, a *Knowledge Repository* is propose which allows us to evaluate different models and web mining studies in order to compare patterns for discovering relevant knowledge from the diverse and changing web data.

This chapter covers the most successful data mining algorithms applied to web data, feature selection and extraction criteria, and model evaluation measures to characterise web mining studies depending on the used technique.

3.1.3 General Terms and Definition of Terms

In the following, the general notation and definition of terms used in this chapter are presented.

- X: Feature space for N objects presented in a dataset, defined by $X = X_{i\,i=1}^{N}$, where each object X_i is defined by n features represented by $X_i = x_{i,j\,j=1}^{n}$.

---

[6] The *Predictive Model Markup Language* (PMML) is an XML standard being developed by the *Data Mining Group* (www.dmg.org)

[7] Application of Data Warehouse architecture to store information about web data.

- Y: Dependent feature for supervised learning problems, where for each object $i \in N$, its value is determined by $Y = y_i|_{i=1}^N$.
- $f(\cdot)$: Theoretical function $f(x)$, $x \in X$ and $f : X \rightarrow Y$, which takes values from the feature space and maps a value in the space of the dependent feature. This function is a theoretical representation of the best approximated function for a given data mining algorithms.
- $h(\cdot)$: Hypothesis function $h(x)$, $x \in X$ and $h : X \rightarrow Y$, with the same behavior than function f. This function represents the empirical estimated function for a given data mining algorithm.
- $\mathcal{M}$: Generic data mining model.

All algorithms and procedures presented through this chapter will be based on this notation, except in case otherwise stated.

3.2   Feature Selection for Web Data

In web mining applications, the number of variables used to characterize Web-objects[8] can be expressed in vectors of hundreds or even thousands of features. This characterization can be influenced by traditional data mining techniques, where feature preprocessing, selection and extraction is evaluated for a given Web-object database. The fact that the Web is a feature-rich environment, which introduces in web mining the curse of dimensionality over the evaluation of different pattern extraction algorithms, from which given the application domain (supervised, unsupervised or incremental learning) the performance and evaluation time of the technique are directly related.

In terms of feature selection and extraction, understood as the construction and selection of useful features to build a good predictor, several potential benefits arises far beyond the regular machine learning point of view for improving the complexity and training capabilities of algorithms. Data visualization and understanding, from an analysts' perspective, is a key factor to determine which algorithms should be considered for the web-data preprocessing. In this context, relevant features for data interpretation could be sub-optimal in terms of building a predictor, as redundant information could be needed for a correct interpretation of a given pattern. Furthermore, in order to determine the set of useful features, the criterion is commonly associated to reducing training time and defy the curse of dimensionality for performance improvements [4, 22, 32].

As a given set of training data can be labeled or unlabelled, some supervised and unsupervised methods for feature selection can be used [62, 68]. Despite the supervised feature selection using an evaluation criteria with respect of the target label, unsupervised feature selection exploits information and patterns extracted from data to determine the relevant set of features.

In this section, Web data pre-processing will be reviewed in terms of feature selection and extraction. Firstly, some of the main feature selection techniques are discussed. Afterwards, some of the main feature extraction methods for Web-mining

---

[8] A Web-object is refered to any data generated over the Web.

are explored. All previous steps for obtaining the correct characterization of web data are covered in other chapters in this book, for different domain tasks, like web-content mining, web-structure mining and web-usage mining.

3.2.1   Feature Selection Techniques

In terms of feature selection, simple measures such as the correlation criteria or information theoretic criteria are well known in many applications of data-mining. Furthermore, meta-algorithms such as Random Sampling, Forward Selection and Backward Elimination are used together with some evaluation measures for determining the most representative set of features, for a given web-mining problem. In the following parts, the main points of these methods will be presented, as well as their most common usage in web-mining applications.

3.2.1.1   Correlation Criteria for Feature Selection

The Pearson correlation coefficient has been widely used in data mining applications [22], and roughly represents the relationship between a given feature and the label, or dependent feature, that is supposed to be predicted.

Despite this measure being widely used by practitioners in several web-mining applications, it can only detect linear dependencies between a given feature and the dependent variable.

3.2.1.2   Information Theoretic Feature Selection

Information theoretic criteria, such as information gain or mutual information, have been used for feature selection and extraction [17, 51], where the amount of information between each feature and the dependent variable is used for ranking in terms of relevance among features. As presented in equation 3.1, the mutual information, which can be considered as a criterion of dependency between the ith feature x_i and the target variable y, is determined by their probability densities $p(x_i)$ and $p(y)$ respectively.

$$I(i) = \int_{x_i} \int_y p(x_i, y) log \frac{p(x_i, y)}{p(x_i)p(y)} dx_i dy \qquad (3.1)$$

In most cases of web-mining applications, web data is represented by discrete amounts of data. Given this, probability distributions are unlikely to be estimated, through which probabilities are determined by frequency tables and counting techniques, as presented by the following expression,

$$I(i) = \sum_{x \in X_i} \sum_{y \in Y} P(x_{i,j}, y) log \frac{P(x, y)}{P(x)P(y)} \qquad (3.2)$$

Using this evaluation criteria, the optimal set of features can be determined by ordering decreasingly given the mutual information (or information gain) parameter, selecting those features which overrides a given threshold.

3.2.1.3 Random Sampling

A simple method for feature selection is to choose features by a uniform random sub-sampling without repeating the whole set of features. This method, known as Random Sampling, has been used when a given dataset is characterized by a high dimensional set of features, where a proportion of features provides enough information on the underlying pattern to be determined. However, if this is not the case, poor results could be obtained as the sub-sampling does not generate a sufficient set of relevant features.

3.2.1.4 Forward Selection and Backward Elimination

Both forward selection and backward elimination meta-algorithms are well known for feature selection in different applications. On the one hand, in the first place forward selection considers an empty set of features are relevant ones are incrementally added. On the other hand, backward elimination considers the deletion of non-relevant features from the whole set of attributes using a given elimination criteria. These methods are used in the scope of wrapper methods [32, 38] and embedded methods [22] for feature selection:

- Wrapper methods: This type of methods requires no prior knowledge of the machine learning algorithm used $\mathcal{M}$, where at each step i of the backward elimination or forward selection process, a subset of features is used for the evaluation of $\mathcal{M}$. Then, the algorithm's performance is evaluated, and finally, from the overall evaluation, the most relevant set of features is determined by the one that maximises the algorithm's performance.
- Embedded methods: For this method, its key component is to use a machine learning algorithm that produces a feature ranking, where the same forward selection or backward elimination procedure described previously, determines the final set of relevant features.

3.2.2 Feature Extraction Techniques

In many Web mining applications, a given set of features can be used to determine a new set for a more effective characterization of the underlying pattern to be extracted. In this sense, methods such as Principal Component Analysis, Singular Value Decomposition, Linear Discriminant Analysis, amongst other feature extraction methods, have been considered in different web-mining and Information Retrieval applications.

3.2.2.1 Principal Component Analysis

In Principal component analysis (PCA) [66] the main idea is to determine a new set of features K from the original feature set X, where $|K| \leq |X| = n$, and each feature $k \in K$ is generated as a uncorrelated (orthogonal) linear combination of the original features. These features are called the principal components of the original set X. In order to determine the new set of features, this method aims at the minimization of the variance of data on each principal components. Given the set of features $X_i \in X$, the objective is to determine the principal component z_k, by estimating the linear combination values $a_k = (a_{1,k}, \ldots, a_{n,k})$, where $z_k = a_k^T \cdot X$

$$\min_{a_k} \quad var(z_k)$$
$$\text{subject to} \quad cov(z_k, z_l) = 0, \; k > l \geq 0 \quad\quad (3.3)$$
$$a_k^T \cdot a_k = 1, a_k \cdot a_l^T = 0, \forall k, l \in K \text{ and } \forall k \neq l$$

where $var(\cdot)$ and $cov(\cdot)$ are the variance function and the covariance functions respectively.

3.2.2.2 Independent Component Analysis

Independent Component Analysis has been used in different web mining applications, such as web usage mining [9], and general web mining feature reduction applications [31]. This method, unlike PCA, aims to determine those features which minimizes mutual information (equation 3.2)between the new feature set Z.

In order to determine the statistical independence between extracted features, the probability distribution function of all features is considered to be determined independently from each feature.

3.2.2.3 Singular Variable Decomposition

As most web data is in fact text, a text specific feature extraction process is introduced. Using the *tf-idf* matrix [46], a Singular Value Decomposition (SVD) of this matrix reduces the dimensions of the term by document space. SVD considers a new representation of the feature space, where the underlying semantic relationship between terms and documents is revealed.

Let matrix X be an $n \times p$ *tf-idf* representation of documents and k an appropiate number for the dimensionality reduction and term projection. Given, $U_k = (u_1, \ldots, u_k)$ an $n \times k$ matrix, the singular values matrix $\mathcal{D}_k = diag(d_1, \ldots, d_k)$, where $\{d_i\}_{i=1}^k$, represents the eigenvalues for XX^T and $V_k = (v_1, \ldots, v_k)$ an $m \times k$ matrix, then the SVD decomposition of X is represented by,

$$X_i = U_i \cdot \mathcal{D}_i \cdot V_i^T \quad\quad (3.4)$$

Here, the expression for $V_i \cdot \mathscr{D}_i$ can be used as a final representation of a given document i. As described in [42], SVD preserves the relative distances in the VSM matrix, while projecting it onto a Semantic Space Model (SSM), which has a lower dimensionality. This allows one to keep the minimum information needed to define the appropiate representation of the dataset.

3.3 Pattern Extraction from Web Data

In this section, main web-mining models for pattern extraction are presented. These models, considered as supervised learning, unsupervised learning and ensemble meta-algorithms will be reviewed, adapted to web mining applications, and their pattern extraction properties.

3.3.1 Supervised Learning Techniques

Over the last years, several supervised algorithms have been developed for previously stated problem tasks, where the main algorithms are Support Vector Machines (SVMs) [6, 48], associated with the regularized risk minimization from the statistical learning theory proposed by Vapnik [53], and the naïve Bayes algorithm [15, 25, 41]. Also, for classification and regression problems, Artificial Neural Networks (ANNs) have been extensively developed since Rosenblatt's perceptron introduced in [44], where the empirical risk minimization is used for adjusting the decision function.

In web mining the target feature can represent different web entities related to classification or regression tasks. For example web-page categorization problems, regression analysis for web user demand forecasting, amongst other applications [34]. Here, supervised learning refers to the fact that the dependent feature $y_j \in Y$ is available. Furthermore, the term learning is related to the fact that the dependent feature must be infered from the set of features X_i, from a training set T and validated from a test set of objects V.

In the following section, these techniques are explained with further details.

3.3.1.1 Support Vector Machines

Support Vector Machines (SVMs) are based on the Structural Risk Minimization (SRM) [6, 53] principle from statistical learning theory. Vapnik stated that the fundamental problem when developing a classification model is not concerning the number of parameters that are and to be estimated, but more, the flexibility of the model, given by *VC-dimension*, introduced by V. Vapnik and A. Chervonenkis [5, 52, 53] to measure the learning capacity of the model. So, a classification model should not be characterized by the number of parameters, but by the flexibility or capacity of the model, which is related to how complicated the model is. The higher the VC-dimension, the more flexible a classifier is.

The main idea of SVMs is to find the optimal hyperplane that separates objects belonging to two classes in a *Feature Space X* ($|X| = n$), maximizing the margin between these classes. The *Feature Space* is considered to be a Hilbert Space defined by a dot product, known as the Kernel function, $k(x,x') := (\phi(x) \cdot \phi(x'))$, where $\phi : \chi \to n$, is the mapping defined that translates an input vector to the *Feature space*. The objective of the SVM algorithm is to find the optimal hyperplane $w^T \cdot x + b$ defined by the following optimization problem,

$$\min_{w,\xi,b} \quad \frac{1}{2}\sum_{i=1}^{n} w_i^2 + C\sum_{i=1}^{N}\xi_i$$
$$\text{subject to} \quad y_i\left(w^T x_i + b\right) \geq 1 - \xi_i \; \forall i \in \{1,..,N\}$$
$$\xi_i \geq 0 \; \forall i \in \{1,..,N\} \tag{3.5}$$

The objective function includes training errors ξ_i while obtaining the maximum margin hyperplane, adjusted by parameter C. Its dual formulation is defined by the following expression, known as the Wolfe dual formulation.

$$\max_{\alpha} \quad \sum_{i=1}^{N}\alpha_i - \frac{1}{2}\sum_{i,j=1}^{N}\alpha_i\alpha_j y_i y_j \cdot k(x_i,x_j)$$
$$\text{subject to} \quad \alpha_i \geq 0, \forall i \in \{1,...,N\} \tag{3.6}$$
$$\sum_{i=1}^{N}\alpha_i y_i = 0$$

Finally, after determining the optimal parameters α in order to classify the labels, the continuous output are represented by,

$$g(x_j) = \sum_{i=1}^{N}\alpha_i y_i \cdot k(x_i,x_j) + b \tag{3.7}$$

The resulting classification for the tth element is given by $h(x_t) = sign(g(X_t))$.

3.3.1.2 Naïve Bayes

Several applications for web mining considered in this model has been proposed [49, 59]. Here, the main objective is to determine the probability that an object composed by the feature set x_i has a given label y_i. In this particular case, it will be considered as a binary classification problem where $y_i = \{+1,-1\}$. For this, using the Bayes theorem, and considering that every feature presented in X_i is independent from each other, the needed probability can be denoted by the following expression,

$$P(y_i|x_i) = \frac{P(y_i) \cdot P(x_i|y_i)}{P(x_i)} = \frac{P(y_i)}{P(x_i)} \cdot \prod_{j=1}^{n} P(x_{ij}|y_i) \tag{3.8}$$

It is easy to prove that considering the conditional probabilities over the labels, the previous equation can be extended to the following expression,

$$ln\left(\frac{P(y_i=+1|x_i)}{P(y_i=-1|x_i)}\right) = ln\left(\frac{P(y_i=+1)}{P(y_i=-1)}\right) + \sum_{j=1}^{n} ln\left(\frac{P(x_{ij}|y_i=+1)}{P(x_{ij}|y_i=-1)}\right) \quad (3.9)$$

And the final decision is given by the following cases, subject to a threshold τ.

$$h(x_i) = \begin{cases} +1 & \text{if } ln\left(\frac{P(y_i=+1|x_i)}{P(y_i=-1|x_i)}\right) > \tau \\ -1 & \text{Otherwise} \end{cases} \quad (3.10)$$

It is important to notice that the previous expression can be determined by continuous updates in the probabilities of the labels (y_i) given the features (x_i),which are determined over the frequencies of the features presented in the incoming stream of messages.

3.3.1.3   Artificial Neural Networks

Artificial Neural Networks (ANN) represents a mathematical model for the operation of biological neurons from brain and, like biological neural structures, ANNs are usually organized in layers. In this context, the neuron is modelled as an activation function that receives stimulus from others neurons, represented by inputs with associate weights $W^1 = (w_1^1,\ldots,w_L^1)$ and $W^2 = (w_1^2,\ldots,w_L^2)$. In a single layer ANN the objective is to minimize the empirical learning risk represented by an error function $\mathcal{E}$, as stated in the following non-linear optimization problem,

$$\min_{W^1,W^2} \quad \mathcal{E} = \sum_{i=1}^{N}\left(y_i - g^1\left(\sum_{l=1}^{L}\sum_{k=1}^{K} w_{l,k}^1 \cdot g^2\left(\sum_{a=1}^{n} w_{a,l}^2 x_a\right)\right)\right) \quad (3.11)$$

Here, the transfer functions $g^1 : L \rightarrow K$ and $g^2 : X \rightarrow L$, where N is the number of objects, L is the number of neurons in the hidden layer, K is the number of labels in the classification problem (for regression problems $K = 1$), and n is the number of features considered for the characterization of objects. This minimization problem is determined by non-linear optimization algorithms, where the most known method to find the set of weights W^1 and W^2 is the back-propagation algorithm [61], in which the convergence towards the minimum error is not guaranteed.

In general, multi-layer ANNs are mainly used in:

- **Classification.** By training an ANN, the output can be used as a feature vector classifier. For example, a web page type (blog, news, magazine, etc.) given web-content features. The input layer receives a n-dimensional vector with page characteristics and in the output layer, the web site can be classified as the type from which it is needed to be classified, depending on the set of neurons result being closer to "1" or "0".

• **Regression.** Given a set of training examples, an ANN can be trained to represent an approximated function on which to model a situation. For example a web site user demand, given the previous historical behaviour of users. When presented with new input examples, the ANN provides the best prediction based on what was learned using the training set of examples.

3.3.2 Unsupervised Techniques

In practice, unsupervised learning in the web-mining context have been widely used as many of the pattern that are to be extracted from web data is from a unsupervised point of view [36, 54, 60, 64, 65]. Many applications on web usage mining determine the web user behaviour [14, 55], and identify both the website keywords [57, 58] and web site key-objects [16].

3.3.2.1 *k*-Means

This algorithm divides a set of data into a predetermined number of clusters [24, 37]. The main idea is to assign each vector to a set of given cluster centroids and then update the centroids given the previously established assignment. This procedure is repeated iteratively until a certain stopping criterion is fulfilled. The number of clusters to be found, k, is a required input value for the k-means algorithm. A set of k vectors are selected from the original data as initial centroids, whose initial values are considered randomly. The clustering process executes an iterative algorithm whose objective is to minimize the overall sum of distances between the centroids and objects, represented as a given stopping criteria in an heuristic algorithm.

It is important to notice that one of the main characteristics of the algorithm is the distance considered between object, which is considered to be relevant in the type of segmentation problem that needs to be solved.

Some clustering evaluation criteria have been introduced for determining the optimal number of clusters [12, 45]. One of the most simple is the Davies-Boulin index [12] (see section 3.4.4). In this case, the main idea is to determine the optimal number of clusters used as stopping rule the minimization of the distance within every cluster and the maximization of the distance between clusters. Likewise, the optimal k-means parameter-finding problem has been extensively discussed in [63].

3.3.2.2 Kohonen Self Organizing Feature Maps

A Kohonen Self Organising Feature Maps (SOFM) [33], is a vector quantization process. This takes a set of vectors as high dimensional inputs and maps them into an ordered sequence. The SOFM maps from the input data space X ($|X| = N$) onto a regular two-dimensional array of nodes or neurons. The output lattice can be rectangular or hexagonal. Each neuron is an N-dimensional vector $m_i \in N$, whose components are the synaptic weights. By construction, all the neurons receive the same input at a given moment of time. This machine learning algorithm is considered as a "non-linear projection of the probability density function of the high dimensional input data onto the bi-dimensional display" [33]. Let $X_i \in X$ be an input data vector. The idea of this learning process is to present X_i to the network and, by using a metric, to determine the most similar neuron (center of excitation, winner neuron).

3.3.2.3   Association Rules

The basic idea in association rules is to find significant correlations among a large data set. A typical example of this technique is the "purchasing analysis", which uses customer buying habits to discover associations among purchased items. In general, the discovered rules are formalized as *if* $< X >$ *then* $< Y >$ expressions.

The formal statement for the described situation is proposed in [1, 2] as follows. Let $I = \{i_1, ..., i_m\}$ be a set of items and $T = \{t_1, ..., t_n\}$ be a set of transactions, where t_i contains a group of items from I. Let $X \subseteq I$ be a group of items from I, a transaction t_i is said to contain X if $X \subseteq t_i$. An association rule is an implication of the form $X \Rightarrow Y$, where $Y \subseteq I$ and $X \cup Y = \emptyset$.

The rule $X \Rightarrow Y$ holds for the transactions set T with support α and confidence β. α is the percentage of transactions in T that contain X and Y. β is the percentage of transactions in T that contain $X \cup Y$. This process can be generalized with the multidimensional association, whose general form is $X_1, ..., X_n \Rightarrow Y$.

3.3.3   Ensemble Meta-algorithms

Ensemble methods have been used to improve results over pattern extraction models. The main idea is to use a large amount of models, along with carefully chosen parameters, combining them to boost a single predictor performance. Ensemble meta-algorithms are generic, and most of data mining models can be used in an ensemble of models, ANNs, SVMs, naïve Bayes and decision Trees, amongst others. Majority voting, boosting, and bagging methods will be reviewed.

3.3.3.1   Majority Voting

This ensemble method is easy to deploy, scale and use in a wide range of applications, associated most of the time in supervised classification algorithms. Consider m different models generated over a training set, and let suppose the simple case for a binary classification problem. After m models are determined, new objects are presented to be classified. Therefore, each object is evaluated for all model, and its results are saved. Afterwards, the predicted label whose frequency at it's highest level, is then considered as the most likely hypothesis for the binary classification problem.

3.3.3.2   Bagging

Bagging (bootstrap aggregation) is considered as a ensemble meta-learning method to improve the predictive accuracy of a given model. The main idea is that given a training set of N objects, the method generates m sets of objects $B_j, j = \{1, ..., m\}$,

where $|B_j| = m < N$. These sets are generated by the bootstrap re-sampling method
with replacements. Each one of the B_j new sets are used to train m models. Results
for these methods can be considered as a regular average over resulting models, or
can be used in a majority voting evaluation schema.

3.3.3.3   Boosting

This meta-learning technique, introduced by Schapire in [47], and further extensions
to the Adaboost meta-algorithm presented by Freund & Schapire in [18, 19] uses a
different re-sampling method than Bagging regarding classification problems. Here,
each subset that evaluates different models is determined initially by using a constant probability over the whole set of objects $\frac{1}{N}$ for each instance. This probability
is updated in time, according to the performance of the models. This technique introduces the concept of weak learners, whose performance is constrained to at least
50% of its accuracy regarding the classification of its performance, and still captures underlying patterns of data. In a nutshell, the idea is that the right set of weak
learners will create a stronger learner.

3.4   Web Mining Model Assessment

In previous sections, several web mining techniques have been introduced. These
learning methods enables the building of hypothesis and models from a set of data.
However, in most cases, an accurate measurement of hypothesis quality is necessary. In fact, there are several methods that measure the model's quality from the
evidence. In the following sections some of these measures will be reviewed.

3.4.1   Evaluation of Classifiers

Given a set of data S determined by objects $X_{i_{i=1}}^N$, an hypothesis function h can be
determined making use of a supervised learner. A simple measure to evaluate the
quality of the classifier h is the *sampling error* defined in equation 3.12.

$$Error_S(h) = \frac{1}{N} \sum_{i=1}^{N} \Delta(f(X_i) \neq h(X_i)) \qquad (3.12)$$

where Δ is a boolean function for which $\Delta(true) = 1$ and $\Delta(false) = 0$. It is relevant to consider an evaluation of the algorithm trained in a different data set, which
was not used for training, given that a biased evaluation and over-fitting decisions
could lead to over-rated predictive models. For a fair evaluation of performance
measures, different approaches have been developed. The most known of these evaluation methods are the Hold-out (train/test) validation and the k-Cross-Validation,
briefly reviewed as follows,

- **Hold Out:** For an evaluation without over-fitting models A more effective approach consists of separating the set S in two subsets: a **Training Set** to define

the hypothesis function h and a **Test Set** to calculate the *sampling error* of the hypothesis.

- **Cross Validation:** k-Cross Validation consists of dividing the evidence in k disjoint subsets of similar size. Then, the hypothesis can be deduced with a set formed by the union of $k - 1$ subsets and the remaining subset is used to calculate the sampling error. This procedure is repeated k times by using a different subset to estimate the sampling error. Then, the final sampling error is the mean of k partial sampling errors. Likewise, the final results are the mean of the experiments with k independent subsets.

3.4.1.1 Evaluation of Hypothesis Based on Cost

Cost sensitive models have been used in several web mining applications. In binary classification tasks all these models are based on the confusion matrix, generated over the different outcomes that a model can have. The main idea is to determine the model that minimizes the overall cost, or maximize the overall gain of the evaluated model for a given application. For the confusion matrix, there are four possible outcomes: Correctly classified type I elements or True Positives (TP), correctly classified type II elements or True Negative (TN), wrongly classified type II elements or False Positive (FP) and wrongly classified type I elements or False Negative (FN). The following evaluation criteria are commonly considered in machine learning and data mining applications.

- The False Positive Rate (FP-Rate) and the False Negative Rate (FN-Rate) as the proportion of wrongly classified type II and type I elements respectively.

$$\text{FP-Rate} = \frac{FP}{FP + TN} \tag{3.13}$$

$$\text{FN-Rate} = \frac{FN}{FN + TN} \tag{3.14}$$

- Precision, that states the degree in which identified as type I elements belongs to the type I elements class. Can be interpreted as the classifier's safety.

$$\text{Precision} = \frac{TP}{TP + FP} \tag{3.15}$$

- Recall, that states the percentage of type I elements that the classifier manages to classify correctly. Can be interpreted as the classifier's effectiveness.

$$\text{Recall} = \frac{TP}{TP + FN} \tag{3.16}$$

- F-measure, the harmonic mean between the precision and recall.

$$\text{F-measure} = \frac{2 * \text{Precision} * \text{Recall}}{\text{Precision} + \text{Recall}} \tag{3.17}$$

- Accuracy, the overall percentage of correct classified elements.

$$\text{Accuracy} = \frac{TP + TN}{TP + TN + FP + FN} \qquad (3.18)$$

- Area Under the Curve (AUC), defined as the area under a ROC curve for the evaluation step

$$AUC = \int_0^1 \frac{TP}{P} d\frac{FP}{N} = \frac{1}{P \cdot N} \int_0^1 TP \, dFP \qquad (3.19)$$

3.4.2 Evaluation of Regression Models

Given an objective theoretical function f, a regression model h and a set of data S formed by N examples, which is one of the most used evaluation measures is the **Mean Squared Error** (MSE):

$$MSE = \frac{1}{N} \sum_{i=1}^{N} (h(X_i) - f(X_i))^2 \qquad (3.20)$$

However, this evaluation criteria does not allow one to calculate the real difference between the estimator and the true value of the quantity being estimated. In fact, MSE has the same unit of measurement as the square of the quantity that is estimated. In an analogy to standard deviation, to fix the units problem, the square root of MSE yields to the **Root Mean Squared Error** (RMSE).

$$RMSE = \sqrt{\frac{1}{N} \sum_{i=1}^{N} (h(X_i) - f(X_i))^2} \qquad (3.21)$$

To square the differences between estimators tends to increase the value of the most extreme errors. This is an issue for MSE and RSE. By using the **Mean Absolute Error** (MAE) this problem is limited.

$$MAE = \frac{1}{N} \sum_{i=1}^{N} |h(X_i) - f(X_i)| \qquad (3.22)$$

Nevertheless, MAE estimates the mean of errors by ignoring its signs. In this sense, relative errors can be used. In other words, by normalising the total squared error by dividing by the total squared error of the simple predictor $\bar{f}$, the **Relative Squared Error** (RSE) can be achieved:

$$RSE = \frac{1}{N} \sum_{i=1}^{1} \frac{(h(X_i) - f(X_i))^2}{(h(X_i) - \bar{f})^2}, \quad \text{where } \bar{f} = \frac{1}{N} \sum_{i=1}^{N} f(X_i) \qquad (3.23)$$

This last measure can be modified in the same way as MSE. Indeed, we could apply square root or use absolute errors over RSE and new measures can be achieved.

3.4.3   MDL Principle

The *Minimum Description Length* (MDL) principle for statistical and machine learning model selection and statistical inference is based on one simple idea: "the best
way to capture regular features in data is to construct a model in a certain class which
permits the shortest description of the data and the model itself". It can be understood as a formalization of **Occam's Razor**, which states that the best hypothesis
for an underlying pattern is the simplest.

Formally, the MDL principle recommends the selection of the hypothesis h that
minimalises the:

$$C(h,D) = K(h) + K(D|h) \qquad (3.24)$$

Where $C(h,D)$ is the complexity in bits of the hypothesis h, $K(h)$ the amount of bits
needed to describe the hypothesis and $K(D|h)$ is the amount required to describe the
evidence D which includes the hypothesis exceptions.

3.4.4   Evaluation of Clustering Models

Unsupervised models are difficult to evaluate because there is not a class or numeric
value from which an error could be estimated. A simple measure, often used to
identify how compact the determined clusters are is presented in equation 3.25.

$$S_{SE}(\mathcal{M}) = \sum_{i=1}^{N} \sum_{k=1}^{K} ||X_i - c_k||^2 \qquad (3.25)$$

where $\mathcal{M}$ is the model formed by K clusters with centroids $c_1, \ldots, c_K$, evaluated
over the data set S with N objects.

Another simple approach, which is often related to analytical purposes of results
obtained, consists of using the distance between clusters. The distance can be defined in different ways:

- **Mean Distance:** It is the mean of the distances between every examples of every
  clusters.
- **Nearest Neighbour Distance:** It is the distance between the nearest neighbours
  of two clusters, i.e., the closest examples.
- **Farthest Neighbour Distance:** It is the distance between the farthest neighbours
  of two clusters, i.e., the more distant examples.

The MDL principle can also be used as an evaluation measure. For example, if using
a particular model, the set E formed by n examples is clustering in k groups, this
model can be used as a codification of set E. Hence, the model which provides the
shortest and more representative codification of E will be the most effective.

In k-means, a frequently used index for determining such number is the Davies-
Bouldin index [12], defined as,

$$\text{Davies-Bouldin} = \frac{1}{K} \sum_{i=1}^{K} \max_{i \; j} \left[\frac{Sk(Q_i) + Sk(Q_j)}{S(Q_i + Q_j)} \right] \qquad (3.26)$$

where K is the number of clusters, Sk refers to the mean distance between objects and their centroid, and $S(Q_i + Q_j)$ is the distance between centroids. This index evaluation states that the best clustering parameter K is obtained from smaller index values.

3.4.5   Evaluating Association Rules

One is used to working with two measures to evaluate the rule quality: **support** and **confidence**. The *support* of a rule of a set of feature values is the number of training database instances (or percentage of training database instances) that contain the values mentioned. While the **confidence** is the number of training database instances (or percentage training database instances) the rule is satisfied.

The traditional task of association rule mining is to find all rules with **high support** and **high confidence**. Nevertheless, in some applications one is interested in rules with *high confidence but low support*. Likewise, some rules with *high support but low confidence* can be used in generalised data.

3.4.6   Other Evaluation Criteria

There are several other criteria to evaluate data mining models, some of which provides interesting insights. However, some of them are considered as subjective and lack useful information in scientific research, but are considered useful in real life applications. The most recurrent criteria are:

- **Interest:** There are models whose assessment measures are great but the knowledge extracted from them is irrelevant. In other words, the model is not interesting for data mining. For example, "if the web user is a mother then she is female". The first ones are related with the previous knowledge of the data miner, if the model accures this knowledge then the model is interesting.
- **Applicability:** A model can be precise regarding training data but it can have poor applicability. For example: "the probability of a web user to become a client depends on his age and gender". However, if we don't rely on with a proactive strategy that provides this web user data (user/password registration, cookie, etc.) then the model is not applicable.

3.5   A Pattern Webhouse Application

In this section, a Pattern Webhouse will be introduced making use of every characteristic and evaluation measure previously presented. For this purpose, Data Webhouse architecture will be presented.

3.5.1   Data Webhouse Overview

The issue of extracting information affects more than the Web environment. Indeed, it is a common problem for any company seeking information from their diary operations systems. At first, companies only created computer systems to register data in order to give support to daily operations. Subsequently, due to market competitiveness, reliable information for decision making was necessary [27]. However, operational systems are not designed for extracting information. In fact, this require complex enquiries whose processing affects the systems performance and the company operations. This situation finally results in time, cost and resources in the area of IT that develop the required reports and decision makers who will not have enough information or material to make their decisions [30].

To handle the aforementioned situation, the Data Warehouse or information repositories appears to supply the information necessities. As described by Inmon in [27], they are "subject oriented, integrated, non-volatile, and time variant collection of data to support management's decisions". It is subject oriented because the data provides non detailed information about business aspects. It is integrated, because it gathers data from different sources and arranges them under a coherent structure. The data is not volatile because it is stable and reliable, and more data can be added. However the data already existing cannot be removed or updated. And furthermore it varies in time because all of the data in the repository is associated with a particular period of time. Kimball in [30] provides a simpler definition: a repository of information is specifically a copy of the transactional data structured for consulting and analysis.

It's worth mentioning that the warehouse information consists of consolidated data that is ready to be consulted. In other words, there was a previous computer process that ran on a schedule of low demands in the systems[9]. Therefore the operational data was converted into information which was stored in the repository [27].

The *Data Web-house* appears as the application of an architecture that has been tested for a decade, the Data Warehouse applied to web data. It was created to assist the information necessities of business with total or partial presence on the Web. Indeed, the proliferation of business based on the Web such as *e-business, e-commerce, e-media* and *e-market* showed how important it was to know more about the Web and its users – all of these under marketing strategies aimed at creating faithful clients.

Applying Data Warehouse to web data brought multiple challenges [29, 54]:

- Web data grows exponentially through time because web logs stores requests each second. Web users change their behaviour in time and web pages are upgraded. Consequently, more data has to be added to the repository.
- The response times expected are much smaller than those in traditional data warehouse. These should be similar to those taken for loading a web page, which is less than 5 seconds.

---

[9] Both operational systems and the repository itself are usually carried at night.

Thanks to Data Webhouse, the web miner, web master or another person interested in statistics about a web site can carry one multidimensional analysis on web data. Indeed, these users see the world in several dimensions, that is, they associate a fact to a group of causal factors [56]. For example, the sale level of a particular product will depend on the price, the period of time and the branch in which it was sold. Consequently, repositories are designed to respond to requirements of information based on how final users perceive the business aspects. In other words, the information is stored based on relevant dimensions for a final user, in a way that facilitates their browsing in the repository in order to enable them to find answers to their enquiries [27].

For example, a commercial manager interested in the amount of visits to the Company web site can look for information about this on each web page, at each hour of each day of the week. According to multidimensional modelling, the fact under scrutiny are the visits, the data, its values, and dimensions associated with them. Also important factors are the web page name, hour and weekday.

3.5.2 About PMML

The Predictive Model Markup Language (PMML) [20, 43] has been developed over the last years for the contribution on predictive analytic models for an analytical infrastructure interoperability. Nowadays, PMML has become a leading standard, whose wide set of features and development has allowed a wide variety of applications to be built in specific industrial applications. The deployment of predictive analytics in the PMML standard has been continuously supported by leading business intelligence providers [67], from which its adoption and growth will be sustainable in the future.

PMML is based on XML standards, where the description of data mining models in this semi-structured language has led to a fluent exchange of models in open standards, contributing to an increase of interoperability for consumers and producers of information and analytical applications. Furthermore, data transformation and preprocessing can be described using PMML models, where output features can be used as components that are served by other PMML applications which represents data mining models.

Today it is straightforward for technological infrastructure components and services using PMML to run on different analytical systems, such as cloud computing architectures. A modeler might use a statistical application to build a model, and the scoring process could be done in a cloud. Also it might be used to produce features from an input data for the statistical model described by the modeler. All this, considering an open standard who is likely to be supported by most data mining and statistical software tools in the future.

3.5.3   Application

The web patterns can be understood as knowledge artefacts which enables one to represent in a compact and semantically rich way large quantities of heterogeneous raw web data. In this sense, since web data may be very heterogeneous, several kinds of patterns exist that can represent hidden knowledge: *Clusters, Association Rules, Regression, Classifiers*, etc. Due to specific characteristics of patterns, specific representation is required for pattern management in order to model, store, retrieve and manipulate patterns in a efficient and effective way.

Given the advantages of PMML to describe data mining models, a repository was proposed that works together with PMML files and describe the different web mining studies. By making use of a Data Warehouse architecture and PMML technology, it is possible to build a repository which represents and stores the knowledge extracted from web data. Generally, *Data Warehouse* and *Web-houses* repositories make use of *star schema* to be implemented in relational databases. Nevertheless, different measures exists for every web mining technique. Hence, more than one fact table is required so a *fact constellation schema* is a more appropriated design. In this sense, the proposed repository is composed by four fact tables and six dimensions.

3.5.3.1   Fact Tables

As shown in figure 3.2, the model includes four fact tables which store measures for each kind of web mining technique. They share seven metrics which are common to every algorithm.

- *Training_time*: It is the amount of time in seconds that the algorithm is trained.
- *Prediction_time*: It is the amount of time in seconds that the algorithm takes to predict a class or numeric value. For some techniques this value will be near to zero.
- *Start_date*: It is the first date in which web data source were taken.
- *End_date*: It is the last date in which web data source were taken.
- *PMML_URL*: It is the path to the PMML file that describe the web mining model.
- *MDL_Hypothesis*: It is the amount of bits that describe the hypothesis or web mining pattern.
- *MDL_Exceptions*: It is the amount of bits that describe the exceptions of hypothesis or web mining pattern.

Likewise, fact tables have particular metrics which are assessment measures whose purpose is to compare different web mining techniques.

1. **Classification_Fact:** Include every measure to assess classifiers. Firstly, the measures based on precision: the *sampling method* used (Training_Test, Cross-Validation and Bootstrapping), the estimated *sampling error* and the *confidence interval* for this last measure. Also, it stores measures based on cost as: *True Positives* (TP), *True Negatives*, *False Positives* (FP), *False Negatives* (FN), *FP Rate*, *FN Rate*, *Precision*, *Recall*, *F-measure*, *Accuracy* and *Area under Curve* (AUC).

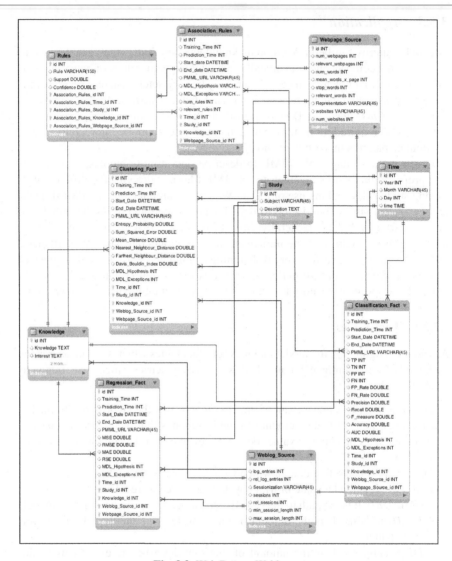

Fig. 3.2 Web Pattern Webhouse

2. **Regression_Fact:** Include the classic measures of quality for regression models. For example: *Mean Squared Error* (MSE), *Root Mean Squared Error* (RMSE), *Mean Absolute Error* (MAE) and *Relative Squared Error* (RSE).
3. **Clustering_Fact:** Different measures are stored in this table. If the clustering model is based on Probability, a probability entropy (S_L) is saved. When models don't provide a probability functions, this table include measures as: *Sum of squared error* per cluster, *Mean distance*, *Nearest neighbour distance* and *Farthest neighbour distance*

4. **AssociationRules_Fact:** The association rule's relevance is given by the support and confidence level. Hence, when we have a web mining study whose purpose is to extract association rules, the *number of rules* and the *number of relevant rules* are stored. Likewise, making use of tables relationship, every extracted rules are stored in another table called **Rules**. Specifically, the *rule* expressed in text and the *support* and *confidence level* associated.

3.5.3.2 Dimensions

Every measure of fact tables is characterised by attributes of five different dimensions.

1. **Time:** This dimension enables on to make tendency analysis regarding metrics included in fact_tables. Every web mining study will be characterised by the date when the algorithm was run, specifically, the *year*, *month*, *day* and *instant* of time.
2. **Study:** Every web mining study has an objective. For which, several algorithms can be used. For example, to study web user behaviour, two clustering techniques can be used: SOFM and K-means. In this case the dimension stores the *subject* and *description* of the study.
3. **Knowledge:** Many times web patterns provide interesting knowledge about the structure, usage and content of a web site. This *knowledge* will be described in text and stored in this dimension. Also, subjective measures such as *Interest* and *Applicability* of the algorithm are stored.
4. **Weblog_Source:** Sometimes a study will make use of web logs as data source. In this case, this dimension characterise the source and the associated preprocessing. Particularly: the *amount of web log entries*, the *number of relevant web log entries* (web log contain irrelevant data), the *sessionization method* used, the *amount of identified sessions*, the *number of relevant sessions* for the study, and finally: the *minimum* and *maximum length* for used sessions.
5. **Webpage_Source:** This dimension characterise the web page as data source and its preprocessing. Particularly, it stores: the *amount of web pages* collected, the *amount of relevant web pages* for the study and the *amount of websites* and *relevant websites* crawled. Regarding the textual information, the *total amount of words*, the *number of stop-words* and the *mean of words per page* are stored. Finally, the *vectorial representation* used is also stored[10].

3.6 Summary

This chapter introduced a specific repository to store *meta-data* about patterns discovered through web mining techniques, In fact, depending on the web mining technique, their parameters, web data quality and pre-process, feature selection and the nature of the web mining study, different patterns can be discovered from web log entries and web page content.

---

[10] For example: Vector Space Model, Boolean Retrieval Model, etc.

Making use of *Data Warehouse* architecture, the same as those commonly known *Data Web-houses*, it is possible to store relevant information about *usage*, *structure* and *content* patterns that enables the manipulation of hidden knowledge.

Firstly, classic web mining techniques are presented, particularly: *association rules*, *classification*, *regression* and *clustering* models. Likewise, the main feature selection techniques applied to web data are introduced.

Later, different assessment measures are presented, specifically, evaluation criteria for the four kinds of web mining models. The idea consists of comparing two studies which uses the same web mining techniques. Also, subjective measures such as *Interest* and *Applicability* are introduced. These last measures enables the rescue of attributes about patterns that previous numeric measures can not.

Finally, the repository is defined as a *fact constellation schema*. Due to the high complexity and nature of patterns and web mining models, it is necessary to divide the information into four fact tables which represents each kind of technique. In this way, fact tables are related to five dimensions: *Time*, *Study*, *Knowledge*, *Weblog_Source* and *Webpage_Source*. On the other hand, the patterns content and their parameters were stored in PMML files, which are a XML standard to diffuse data mining models. This technology allows one to reduce the repository complexity, particularly, the granularity of the information. Hence, the proposed repository and PMML files work effectively in managing the knowledge contained in patterns discovered from web data.

Acknowledgements. This work was supported partially by the FONDEF project DO8I-1015 and the *Web Intelligence Research Group* (wi.dii.uchile.cl) is greatly acknowledge.

References

1. Agrawal, R., Imieliński, T., Swami, A.: Mining association rules between sets of items in large databases. SIGMOD Rec. 22(2), 207–216 (1993)
2. Agrawal, R., Srikant, R.: Fast algorithms for mining association rules in large databases. In: VLDB 1994: Proceedings of the 20th International Conference on Very Large Data Bases, pp. 487–499. Morgan Kaufmann, San Francisco (1994)
3. Arnold, W.A., Bowie, J.S.: Artificial Intelligence: A Personal Commonsense Journey. Prentice-Hall, Englewood Cliffs (1985)
4. Blum, A.L., Langley, P.: Selection of relevant features and examples in machine learning. Artif. Intell. 97(1-2), 245–271 (1997)
5. Blumer, A., Ehrenfeucht, A., Haussler, D., Warmuth, M.K.: Learnability and the vapnik-chervonenkis dimension. J. ACM 36(4), 929–965 (1989)
6. Boser, B.E., Guyon, I.M., Vapnik, V.N.: A training algorithm for optimal margin classifiers. In: COLT 1992: Proceedings of the fifth annual workshop on Computational learning theory, pp. 144–152. ACM Press, New York (1992)
7. Catania, B., Maddalena, A.: Hershey, PA, USA
8. Catania, B., Maddalena, A., Mazza, M.: Psycho: A prototype system for pattern management. In: Böhm, K., Jensen, C.S., Haas, L.M., Kersten, M.L., Larson, P.-., Ooi, B.C. (eds.) VLDB, pp. 1346–1349. ACM, New York (2005)

9. Chimphlee, S., Salim, N., Ngadiman, M.S.B., Chimphlee, W., Srinoy, S.: Independent component analysis and rough fuzzy based approach to web usage mining. In: AIA 2006: Proceedings of the 24th IASTED international conference on Artificial intelligence and applications, pp. 422–427. ACTA Press, Anaheim (2006)
10. Cooley, R., Mobasher, B., Srivastava, J.: Data preparation for mining world wide web browsing patterns. Knowledge and Information Systems 1, 5–32 (1999)
11. Davenport, T., Prusak, L.: Working Knowledge: How Organizations Manage What They Know. Harvard Business School Press, Cambridge (1997)
12. Davies, D.L., Bouldin, D.W.: A cluster separation measure. IEEE Transactions on Pattern Analysis and Machine Intelligence 1, 224–227 (1979)
13. Davis, R., Shrobe, H., Szolovits, P.: What is knowledge representation. AI Magazine 14(1), 17–33 (1993)
14. Dell, R.F., Román, P.E., Velásquez, J.D.: Web user session reconstruction using integer programming. In: Web Intelligence, pp. 385–388. IEEE, Los Alamitos (2008)
15. Domingos, P., Pazzani, M., Provan, G.: On the optimality of the simple bayesian classifier under zero-one loss. Machine Learning, 103–130 (1997)
16. Dujovne, L.E., Velásquez, J.D.: Design and implementation of a methodology for identifying website keyobjects. In: Velásquez, J.D., Ríos, S.A., Howlett, R.J., Jain, L.C. (eds.) Knowledge-Based and Intelligent Information and Engineering Systems. LNCS, vol. 5711, pp. 301–308. Springer, Heidelberg (2009)
17. Fleuret, F.: Fast binary feature selection with conditional mutual information. Journal of Machine Learning Research 5, 1531–1555 (2004)
18. Freund, Y., Schapire, R.E.: A decision-theoretic generalization of on-line learning and an application to boosting. In: Vitányi, P.M.B. (ed.) EuroCOLT 1995. LNCS, vol. 904, pp. 23–37. Springer, Heidelberg (1995)
19. Freund, Y., Schapire, R.E.: Experiments with a new boosting algorithm. In: ICML, pp. 148–156 (1996)
20. Grossman, R.L.: What is analytic infrastructure and why should you care? SIGKDD Explor. Newsl. 11(1), 5–9 (2009)
21. Grossman, R.L., Hornick, M.F., Meyer, G.: Data mining standards initiatives. Commun. ACM 45(8), 59–61 (2002)
22. Guyon, I., Elisseeff, A.: An introduction to variable and feature selection. Journal of Machine Learning Research 3, 1157–1182 (2003)
23. Hah, J., Fu, Y., Wang, W., Koperski, K., Zaiane, O.: Dmql: A data mining query language for relational databases (1996)
24. Hartigan, J.A., Wong, M.A.: A K-means clustering algorithm. Applied Statistics 28, 100–108 (1979)
25. Hastie, T., Tibshirani, R., Friedman, J.: The Elements of Statistical Learning: Data Mining, Inference, and Prediction, 2nd edn. Springer Series in Statistics. Springer, Heidelberg (2009); corr. 3rd printing edition (September 2009)
26. Imieliński, T., Virmani, A.: Msql: A query language for database mining. Data Min. Knowl. Discov. 3(4), 373–408 (1999)
27. Inmon, W.H.: Building the Data Warehouse, 4th edn. Wiley Publishing, Chichester (2005)
28. Elder IV, J.F., Fogelman-Soulié, F., Flach, P.A., Zaki, M.J. (eds.): Proceedings of the 15th ACM SIGKDD International Conference on Knowledge Discovery and Data Mining, Paris, France, June 28 - July 1, 2009. ACM, New York (2009)
29. Kimball, R., Merx, R.: The Data Webhouse Toolkit. Wiley Computer Publisher, Chichester (2000)

30. Kimball, R., Ross, M.: The Data Warehouse Toolkit: The Complete Guide to Dimensional Modeling, 2nd edn. Wiley, Chichester (2002)
31. Klopotek, M.A., Wierzchon, S.T., Trojanowski, K.: Intelligent Information Processing and Web Mining: Proceedings of the International IIS: IIPWM 2006 Conference, Ustron, Poland, June 19-22, 2006. Advances in Soft Computing. Springer-Verlag New York, Inc., Secaucus (2006)
32. Kohavi, R., John, G.H.: Wrappers for feature subset selection. Artif. Intell. 97(1-2), 273–324 (1997)
33. Kohonen, T., Schroeder, M.R., Huang, T.S. (eds.): Self-Organizing Maps. Springer-Verlag New York, Inc., Secaucus (2001)
34. Larsen, J., Hansen, L.K., Have, A.S., Christiansen, T., Kolenda, T.: Webmining: learning from the world wide web. Computational Statistics & Data Analysis 38(4), 517–532 (2002)
35. Liu, B.: Web Data Mining: Exploring Hyperlinks, Content and Usage Data, 1st edn. Springer, Heidelberg (2007)
36. Luo, P., Lin, F., Xiong, Y., Zhao, Y., Shi, Z.: Towards combining web classification and web information extraction: a case study. In: IV et al.: [28], pp. 1235–1244
37. MacQueen, J.B.: Some methods for classification and analysis of multivariate observations. In: Le Cam, L.M., Neyman, J. (eds.) Proc. of the fifth Berkeley Symposium on Mathematical Statistics and Probability, vol. 1, pp. 281–297. University of California Press (1967)
38. Maldonado, S., Weber, R.: A wrapper method for feature selection using support vector machines. Inf. Sci. 179(13), 2208–2217 (2009)
39. Markov, Z., Larose, D.T.: Data Mining the Web: Uncovering Patterns in Web Content, Structure, and Usage. Wiley Interscience, Hoboken (2007)
40. Meo, R., Psaila, G., Ceri, S.: An extension to sql for mining association rules. Data Min. Knowl. Discov. 2(2), 195–224 (1998)
41. Mitchell, T.M.: Machine Learning. McGraw-Hill, New York (1997)
42. Papadimitriou, C.H., Tamaki, H., Raghavan, P., Vempala, S.: Latent semantic indexing: a probabilistic analysis. In: PODS 1998: Proceedings of the seventeenth ACM SIGACT-SIGMOD-SIGART symposium on Principles of database systems, pp. 159–168. ACM Press, New York (1998)
43. Pechter, R.: What's pmml and what's new in pmml 4.0? SIGKDD Explor. Newsl. 11(1), 19–25 (2009)
44. Rosenblatt, F.: Principles of Neurodynamics: Perceptrons and the Theory of Brain Mechanisms. Spartan Books (1962)
45. Rousseeuw, P.: Silhouettes: a graphical aid to the interpretation and validation of cluster analysis. J. Comput. Appl. Math. 20(1), 53–65 (1987)
46. Salton, G., Wong, A., Yang, C.S.: A vector space model for automatic indexing. Commun. ACM 18(11), 613–620 (1975)
47. Schapire, R.E.: The strength of weak learnability. Mach. Learn. 5(2), 197–227 (1990)
48. Schölkopf, B., Smola, A.J.: Learning with Kernels: Support Vector Machines, Regularization, Optimization, and Beyond. MIT Press, Cambridge (2001)
49. Sebastiani, F.: Text categorization. In: Zanasi, A. (ed.) Text Mining and its Applications to Intelligence, CRM and Knowledge Management, pp. 109–129. WIT Press, Southampton (2005)
50. Terrovitis, M., Vassiliadis, P., Skiadopoulos, S., Bertino, E., Catania, B., Maddalena, A.: Modeling and language support for the management of pattern-bases. In: International Conference on Scientific and Statistical Database Management, vol. 0, p. 265 (2004)

51. Torkkola, K.: Feature extraction by non parametric mutual information maximization. Journal of Machine Learning Research 3, 1415–1438 (2003)
52. Vapnik, V., Chervonenkis, A.: On the uniform convergence of relative frequencies of events to their probabilities. Theory of Probability and its Applications 16, 264–280 (1971)
53. Vapnik, V.N.: The Nature of Statistical Learning Theory (Information Science and Statistics). Springer, Heidelberg (1999)
54. Velasquez, J.D., Palade, V.: Adaptive Web Sites: A Knowledge Extraction from Web Data Approach. IOS Press, Amsterdam (2008)
55. Velasquez, J.D., Palade, V.: Building a knowledge base for implementing a web-based computerized recommendation system. International Journal of Artificial Intelligence Tools 16(5), 793–828 (2007)
56. Velásquez, J.D., Palade, V.: A knowledge base for the maintenance of knowledge extracted from web data. Knowledge Based Systems 20(3), 238–248 (2007)
57. Velasquez, J.D., Yasuda, H., Aoki, T., Weber, R.: A new similarity measure to understand visitor behavior in a web site. IEICE Transactions on Information and Systems, Special Issues in Information Processing Technology for web utilization E87-D(2), 389–396 (2004)
58. Velasquez, J.D., Rios, S.A., Bassi, A., Yasuda, H., Aoki, T.: Towards the identification of keywords in the web site text content: A methodological approach. International Journal of Web Information Systems information 1(1), 53–57 (2005)
59. Wang, Y., Hodges, J., Tang, B.: Classification of web documents using a naive bayes method. In: ICTAI 2003: Proceedings of the 15th IEEE International Conference on Tools with Artificial Intelligence, p. 560. IEEE Computer Society Press, Washington (2003)
60. Wen, C.W., Liu, H., Wen, W.X., Zheng, J.: A distributed hierarchical clustering system for web mining. In: Wang, X.S., Yu, G., Lu, H. (eds.) WAIM 2001. LNCS, vol. 2118, pp. 103–113. Springer, Heidelberg (2001)
61. Werbos, P.J.: The roots of backpropagation: from ordered derivatives to neural networks and political forecasting. Wiley Interscience, New York (1994)
62. Wolf, L., Shashua, A.: Feature selection for unsupervised and supervised inference: The emergence of sparsity in a weight-based approach. J. Mach. Learn. Res. 6, 1855–1887 (2005)
63. Wu, J., Xiong, H., Chen, J.: Adapting the right measures for k-means clustering. In: IV et al.: [28], pp. 877–886
64. Xu, R., Wunsch, I.: Survey of clustering algorithms. IEEE Transactions on Neural Networks 16(3), 645–678 (2005)
65. Yin, Z., Li, R., Mei, Q., Han, J.: Exploring social tagging graph for web object classification. In: IV et al.: [28], pp. 957–966
66. Young, T.Y.: The reliability of linear feature extractors. IEEE Transactions on Computers 20(9), 967–971 (1971)
67. Zeller, M., Grossman, R., Lingenfelder, C., Berthold, M.R., Marcade, E., Pechter, R., Hoskins, M., Thompson, W., Holada, R.: Open standards and cloud computing: Kdd-2009 panel report. In: KDD 2009: Proceedings of the 15th ACM SIGKDD international conference on Knowledge discovery and data mining, pp. 11–18. ACM Press, New York (2009)
68. Zhao, Z., Liu, H.: Spectral feature selection for supervised and unsupervised learning. In: ICML 2007: Proceedings of the 24th international conference on Machine learning, pp. 1151–1157. ACM Press, New York (2007)

Chapter 4
Web Content Mining Using MicroGenres

Václav Snášel, Miloš Kudělka, and Zdeněk Horák

Abstract. The size and growth of the current Web is still creating new challenges to researchers. For example, one of these challenges is the improvement of user familarity to a large number of Web pages. Today's search engines provide tools that allow users to refine their queries. One way is the refinement of a query based on the analysis of web content. Possible outcomes are not only recommended collocations, but also recommended page genres (e.g., discussion forums, etc.). It is proving to be very useful to provide the details of page content when viewing the page. Not only text snippets, but also parts of the page menu, for certain pages how many posts are present in the discussion, what day the review was created, or what the price is of a product sold on the page. Obtaining this information from unstructured or semi-structured content is not straightforward. In this chapter the development of methods capable of detecting and extracting information from Web pages will be addressed. The concept of objects, called MicroGenre will be presented. Finally we also present experiments with our own Pattrio method, which provides a way to detect objects placed on Web pages.

4.1 Introduction

The Web is like a big city. It has its center, periphery, buildings and other facilities with various purposes, and communications that connect one another. Also people live in the city, and people move city over time. In order to study the city, one must tackle many tasks. It would be a mistake, however, to study the city without people. Certainly it could be achieved if one moves all the people away from the city, but it would be difficult to find the purpose and meaning of many things. The comparison

Václav Snášel · Miloš Kudělka · Zdeněk Horák
Faculty of Electrical Engineering and Computer Science,
VŠB–Technical University of Ostrava, Czech Republic
e-mail: vaclav.snasel@vsb.cz, kudelka@vsb.cz,
     zdenek.horak.st4@vsb.cz

J.D. Velásquez and L.C. Jain (Eds.): Advanced Techniques in Web Intelligence – 1, SCI 311, pp. 79–111.
springerlink.com   © Springer-Verlag Berlin Heidelberg 2010

of the Web and a city is not self-serving. Many of the previous approaches and methods for analyzing Web content resemble the analysis of an vacant city. For example one can consider approaches dealing with the analysis of the text part of a Web page regardless of its structure, visual appearance or development through time. On the other hand one can consider approaches dealing with the structure of the page with little or no emphasis on the content of particular parts of the Web page.

It was only a matter of time when aspects based on human activity would perform an increasing role in the analysis of Web content. However the analysis of Web content is very complex now. It embraces both technical aspects (starting with HTML code, continuing through the overall structure of the page, finishing with the visual representation of the page) and the human factor, which answers the question: "How do the people do it? How do they use it?". Therefore, in the first part of this chapter the areas of Web usability and Web design patterns will be addressed. In our opinion, these two areas are not connected to the analysis of Web content very often. Regardless of this, they can provide many interesting ideas that may help to improved one's understand of how to proceed with the analysis.

This chapter is organized as follows. In Sect. 4.2, a meaning of Web content mining and typical tasks are explained. In Sect. 4.3, basic principles concerning Web usability and Web design patterns are described. Sect. 4.4 is devoted to the survey of recent approaches. In Sect. 4.5 the term *MicroGenre* as a building block of Web page is defined. Our own Pattrio method, which focusses on the detection of MicroGenres will presented. In Sect. 4.6, experiments related to the successfulness of this method's usability will be described. The last section of the chapter is devoted to a summary and prospects for further research.

4.2   Web Content Mining Summary

The current World Wide Web is the result of interaction between authors of ideas and users. This interaction is permanent and each of these groups takes part in the future direction of the Web. One of the key elements of this progress is the view of the Web from the opposing side - from the side of the data that results from this permanent interaction. Automatically obtained data can be used for various tasks, especially tasks connected with information retrieval. The approaches known from the field of data mining (see Han and Kamber [30]) are applied in the Web environment during the Web evolution and therefore new specific approaches arrise. These approaches define the field of Web mining.

Web mining is the usage of data mining technology on the Web (see Han and Chang [32]). Specifically, it is the case of finding and extracting information from sources that relate to the user's interaction with Web pages. In 2002 (see Han and Chang [32]), several challenges had been formulated for developing an intelligent Web: Web page complexity far exceeds the complexity of traditional text documents, the Web constitutes highly dynamic information, the Web serves as a broad spectrum for user communities, and only a small portion of Web pages contain

truly relevant or useful information sources of any traditional text document collection. Therefore, the following issues should be solved primarily: mining Web
search-engine data, analyzing Web link structures, classifying Web documents automatically, mining Web page semantic structures and page contents, mining Web
dynamics, building a multilayered, multidimensional Web, etc.

Most tasks are still relevant today. In [40], Kosala and Blockeel define three areas
of Web mining: Web content mining, Web structure mining and Web usage mining.

Web content mining: It describes the discovery of useful information from the
    Web contents, data, and documents. The Web content data consists of unstruc-
    tured data such as free texts, semi-structured data such as HTML documents, and
    more structured data such as data in tables or database generated HTML pages.
Web structure mining: This attempts to discover the model underlying the linking
    structures of the Web.
Web usage mining: This attempts to make sense of the data generated by the Web
    user's sessions or behaviors.

Our research focuses on Web content mining whose purpose is to analyze Web pages
in order to find out which useful information is included in the Web page (from the
user's point of view). In this area, two sub-divisions can be found. The first one
is based on information retrieval and its purpose is to find information useful for
locating relevant Web pages in large collections (Web searching). The other one is
based on information extraction and its purpose (see Chang et al. [11]) is to find
structural information that can be, for example, saved in the database and to process
it accordingly (for example the name and price of products).

Liu and Chang [49] consider twelve topics on Web content mining. Following
this, there is a classification of mining tasks:

Structured Data Extraction: One of the reasons for the importance of this topic is
    that structured data on the Web is often very important as they represent essential
    information on their host pages (e.g., lists of products and services).
Unstructured Text Extraction: Most Web pages can be seen as text documents.
    This research is closely related to text mining, information retrieval and natural
    language processing.
Web Information Integration: Web sites may use different syntaxes to express sim-
    ilar or related information. In order to make use of or extract information from
    multiple sites to provide value added services, it is necessary to semantically
    integrate information from multiple sources.
Building Concept Hierarchies: Because of the huge size of the Web, the organiza-
    tion of information is obviously an important issue. A linear list of ranked pages
    is usually produced by search engines. The standard method for information or-
    ganization is through concept hierarchy or categorization. Popular techniques
    include clustering methods.
Segmenting Web Pages & Detecting Noise: A typical Web page consists of many
    blocks or areas (main content areas, navigation areas, advertisements, etc). It is
    useful to separate these areas automatically for several practical applications.

Mining Web Opinion Sources: Companies usually conduct customer surveys or engage external consultants to find such reviews about their products. There are numerous Web sites and pages containing customer opinions (e.g., customer reviews of products, discussion groups, and blogs).

Human factor plays a key role in all the aforementioned tasks. These are still the same tasks on an abstract level, however as the requirements and technology evolve, also the implementation environment changes. In order to successfully perform these tasks the evolution of the Web and the way users work must be followed. One must to adapt methods from the area of Web content mining. The field which formalizes the interaction in the Web environment is the field of Web usability.

4.3   Web Usability Basics

Web usability is closely linked to *User Centered Design* (UCD, see [75]). In a wider sense, UCD is a philosophy which results from the process of software system development. In this system, the user is involved in the first stage of each phase. The main difference from other approaches is that UCD tries to optimize the user interface, so that it:

- Corresponds to what users are used to.
- Does not make the user change their way of working.

From the user's point of view the result of the development process in the Internet environment is a Web page. Each page is prepared for the user and it is dependent on its quality if the user gains what he needs from interacting with it. The problem is that the quality is very difficult to measure in advance. Users have to contribute to the verification of Web page quality. If the Web page should be one of high quality, it should fulfill certain principles. Nielsen defines usability as the following (see [55]):

Usability is a quality attribute that assesses how easy user interfaces are to use. Usability is defined by five quality components:

1. Learnability: How easy is it for users to accomplish basic tasks the first time they encounter the design?
2. Efficiency: Once users have learned the design, how quickly can they perform the tasks?
3. Memorability: When users return to the design after a period of absence, how easily can they reestablish proficiency?
4. Errors: How many errors do users make, how severe are these errors, and how easily can they recover from the errors?
5. Satisfaction: How pleasant is it to use the design?

In the book [56], Nielsen and Loranger present many tests with users and many important recommendations for Web page creators emerges from the results of these tests. One of which is the definition of what Web page authors should accept and avoid.

Standard: Eighty percent or more of Web sites use the same design approach. Users strongly expect standard elements to work a certain way when they visit a new site because that is how things almost always work (the standard can be understood as a formal or universally acceptable norm regardless of the user context.).

Convention: About 50 to 79 percent of Web sites use the same design approach. Users expect conventional elements to work a certain way when they visit a new site because that is how it usually works (the convention can be understood as a custom in a specific context, e.g. in the discussion forum.).

Confusion: Withrespect to these elements, no single design approach dominates, and even the most popular approach is used by less than half of Web sites. For such design elements, users do not know what to expect when they visit a new site.

Regarding the fact that the recommendations coming out of Web usability are acknowledged by Web authors, one can, on the other hand, focus on certain features (from the point of view of Web content mining) which result from standards and conventions.

However, one of the problems is that recommendations in the field of Web usability are not completely formalized. In [28], Graham concludes that recommendations resulting from Web usability can be formalized with the use of patterns (for more about patterns see section 4.3.1). One of the characteristics of the pattern which describes verified experience is the concise name, which characterize the solved task.

The methodology of using patterns for Web design is very thoroughly elaborated in [74] (Van Duyne et al.). Classification of patterns and important recommendations for developers that can be found link not only to Web usability.

Further interesting pieces of information related to the usability view on the Web page which are stated in [72] (Tidwell). These pieces of information are related to page layout and parts of page perception. The theory of grouping and alignment was developed early in the 20th century. The Gestalt psychologist described several layout properties . Some of which seem to be important for our visual systems (see Figure 4.1):

Proximity   occurs when elements are placed close together. These elements tend to be perceived as a group. If things are close together viewers will associate them with one another. This is the basis for a strong grouping of content and controls on a user interface.

Similarity   occurs when elements look similar to one another. These elements tend to be perceived as a group. If two items are the same shape, size, color, or orientation, then viewers will also associate them with each other.

Continuity   occurs when the eye is forced to move through one element and continue through to another element. Our eyes want to see continuous lines and curves formed by the alignment of smaller elements.

Closure   occurs when elements are not completely enclosed in a space. If enough of the information is indicated, these elements tend to be perceived as a group, the missing elements are filled in automatically: One also expects to see

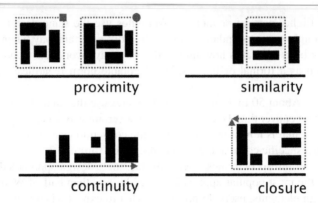

Fig. 4.1 Gestalt Principles (Proximity, Similarity, Continuity, Closure)

simple closed forms, like rectangles and blobs of whitespace that are not explicitly drawn for us.

In paper [23], Flieder and Mödritscher propose a set of principles based on Gestalts in a pattern language form. At the lowest level, these principles follow Gestalt principles and describe the essential syntax, which establishes the rules for combining words, shapes, and images. These morphological elements have various visible properties such as color, size, thickness, texture, orientation, and transparency.

As aforementioned, it is difficult to measure the quality of a Web page exactly. However, Ivory et al. describe some characteristics which are unsuitable for Web page quality evaluation (see [35]). Selected results are listed in Table 4.1.

Table 4.1 Selected Web page design metrics

Metric	Description	Significance
Word Count	Total words on a page	0.150
Body Text %	Percentage of words that are body vs. headers	0.824
Emphasized Body Text %	Portion of body text that is emphasized	0.672
Text Positioning Count	Changes in text position from flush left	0.244
Text Cluster Count	Text areas highlighted (color, regions,...)	0.000
Link Count	Total links on a page	0.000
Page Size	Total bytes for the page and images	0.001
Graphics Count	Total images on a page	0.002
Color Count	Total colors employed	0.001
Font Count	Total font face and size combinations employed	0.836

On the basis of these characteristics, the authors performed experiments with the involvement of users, searching for links between users' evaluations of Web pages and the characteristics. In paper [34], Ivory and Megraw work with these characteristics as well as with Web design patterns and watch over time how the patterns develop. The conclusions which emerged from these experiments with users and through pattern evolution are linked to Web usability. Moreover they provide useful technical basics for procedures linked to Web content mining.

Also eye-tracking is an important field in user interface design (see Goldberg et al. [27]). It is about how a user's goal influences the way they read and traverse a Web page, which parts of a page users attend to first, how people react to advertising, where they look first for common page elements, how they respond to text, pictures, and much more. In [24], Gagneux et al. formulate an answer to the question *"In which way does the visual organization of the Web pages help to lead the visual exploration for an information retrieval?"*. A specific goal can be described by respecting two characteristics:

• It must be compatible with the set of the designer's intentions.
• It must be compatible with the set of the user's potentials.

As a result, methods of Web page analysis have to be based on the relations among human perception, cognitive sciences, and biology. The Web page is usually split up into two common structures:

• The physical structure that expresses the organization as geometric blocks created with homogeneous characteristics. This is the layout of the document.
• The logical structure that expresses the semantic description of the physical organization, which deals with the human interpretation of each of the blocks (line, section, title, paragraph...).

Based on these characteristics, good Web pages are characterized by a simple structure, a basic scan path, a low number of fixations, a heterogeneous distribution of fixations and an understandable physical structure on a low level (during the pre-attentive glance). In contrast, bad quality Web pages are characterized by many fixations, a homogeneous distribution of fixations on images, a rather high density of information and an absense of logical path on the physical structure.

Simply said, the more the developers keep to the mentioned standards, agreements, principles and patterns, the more usable the Web page is. Then it can be expected from the other side that the more usable the page is, the more synoptic it is also for those who deal with algorithm design for its automatic analysis.

The persistent problem is the low level of formalization of methods, standards and conventions. As mentioned above, one possible way to formalize such things is the design patterns. Using this formalization, we can - at a certain level of granularity - describe a Web page.

4.3.1 Web Design Pattern Basics

Design patterns and pattern languages came from the architectural of Christopher Alexander and his colleagues. From the mid sixties to the mid seventies, Alexander and his colleagues defined a new approach to architectural design. The new approach centered on the concept of pattern languages, and is described in a series of books (see [1, 2]).

Alexander's definition of pattern is as follows (see [1]): "*Each pattern describes a problem which occurs over and over again in our environment, and then describes the core of the solution to that problem, in such a way that you can use this solution a million times over, without ever doing it the same way twice*".

According to [72] patterns are structural and behavioral features improve the applicability of software architecture, a user interface, a web site or something else in some domain. They make things more usable and easier to understand. Patterns are descriptions of the best practices within a given design domain. They capture common and widely accepted solutions, their validity is empirically proved. Patterns are not novel ideas, they are captured experiences and each of their implementations is a little different.

In software architecture, design patterns are a mechanism for expressing object-oriented design experience (see [25]). Design patterns identify, name, and abstract common themes in object-oriented design. They provide a common vocabulary and reduce system complexity by naming and defining abstractions. Design patterns can be considered reusable micro-architecture of overall system architecture.

Interaction design patterns supply a solution to typical problems within the design of the user interface. A typical example is the organization of user controls into lists or tabs and so on. Interaction design patterns describe on a general level how to make structures from information within user interface. They also tell which components to use, how they should work together and how to work with them (see Tidwell [72]).

A good example is also the Web design patterns. These are design patterns related to the web. A typical example of a Web design pattern can be the Forum pattern (see Figure 4.2). This pattern is meant for designers who need to implement this element on an independent web page or as part of another web page. The pattern describes key solution features without implementation details.

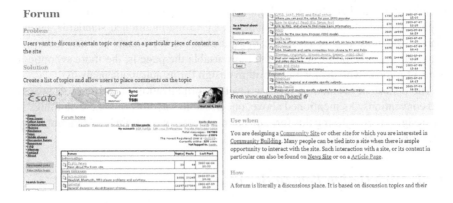

Fig. 4.2 Sample of Forum pattern (www.welie.com)

Pattern Language

An important aspect about patterns is that they are usually related to each other and occur in groups. Another key feature of patterns is that they can be worked within a fashion similar to that of a dictionary, because each pattern has its name, which characterizes its use. It is therefore possible to create so called pattern languages (see Dearden and Finlay [16]), which are groups of patterns covering a certain domain. When using the patterns, neither the author nor the user have to think about how to describe what should be on a Web page in detail. He/she can use the dictionary of pattern language. Pattern languages therefore provide a comprehensive language with which we are able to communicate with the user about what is expected on a Web page.

Borchers formally describes the pattern language in the following way (see [6]):

Definition 4.1. A Pattern Language is a directed acyclic graph $PL = (P,R)$ with nodes $P = \{p_1,\ldots,p_n\}$ and edges $R = \{r_1,\ldots,r_m\}$.

1. Each node $p_i \in P$ is called a pattern.
2. For $p_i, p_j \in P : p_i$ references p_j if exists $r_k = (p_i,p_j) \in R$.
3. The set of edges leave a node p_i, which is called its references.
4. The set of edges entered p_i, is called its context.

The definition implies that there can be relations among the patterns, which can be described on an even more detailed level. Relationships between patterns can be defined by some levels. In [77], Welie uses a simple classification.

Aggregation: This is a form of a "has-a" relationship.

Specialization: This is a "is-a" relationship, one pattern is a more specific version of another pattern.

Association: This is not a "has-a" or "is-a" relationship but simply a "related-to" relationship. A pattern may be associated with other patterns because they also often occur in the larger context of the design problem, or because the patterns are alternatives for the same kind of problems.

Patterns are simple descriptions of repeated problems and their solutions. They are intended for developers and they do not contain technical details. That is why their instances in real solutions are not stereotypes and in many cases they are difficult to distinguish. Besides that they deal with different levels of detail (see Welie [78]). It can be a level of a whole page, and also that, of a small part of a page, e.g. user login. On the other hand, some parts of Web pages can be so basic that it is not necessary to describe them in the form of patterns. However they can be interesting from a page description point of view.

4.4   Recent Methods

Firstly, we will describe approaches and methods, which are used for the analysis of Web page structure and Web page segmentation (see [48]). Methods can be classified in several ways:

1. On the basis of role, which the DOM tree plays on DOM based and visual layout based (see [7, 15]). The first category is especially focused on the detection of interesting sub-trees in the DOM tree. The second category uses DOM only to a limited extent (or they are not used at all) and in order to ensure the structure of a page is from the point of view of external appearances. DOM is W3C's Document Object Model is a standard application programming interface to the structure of documents – see [76].

2. On the basis of human participation in the analysis (see [48]) there are supervised (wrapper induction, e.g. [3, 10]) and unsupervised (automatic extraction) methods.

3. On a basis of the fact, if the analysis assumption is a known Web page structure (see e.g. [88, 90]). These are template-dependent and template-independent methods.

Currently the trend is evolving towards automatic, template-independent and visual layout based approaches. That is why in this text only selected approaches in this field is mentioned.

Chen et al. describe in [12] the function-based Object Model (FOM). This approach is based on understanding authors' intentions, not on understanding semantic Web pages. Every object on a Web page serves certain functions (basic and specific Functions) which reflect the authors' intentions towards the purpose of an object. The FOM model for Web page understanding is based on this consideration. FOM includes two complementary parts: basic FOM (BFOM) based on the basic functional properties of an object, and specific FOM (SFOM) based on the category of an object. It is supposed that by combining BFOM and SFOM, a thorough understanding of the authors' intentions of creating a Web page can be determined.

In [81], Yang et al. present an approach of automatic analysis of the semantic structure of HTML pages based on detecting visual similarities of content objects on Web pages. Effective content organization is an essential factor of high quality content services. The content is divided into categories and each holds records of related subtitles. Records in one category are normally organized in ways which have a consistent visual layout style. Boundaries between different categories are marked with a number of visual styles or separators. They define five types of page objects (simple objects, container objects, group objects, list objects, and structured documents). They detect these objects and measure their similarity by fuzzy comparison rules.

Kovacevic et al. present in [41] techniques based on a virtual screen defines a coordinate system for specifying the position of objects inside Web pages. Authors use a specific format of the page for rendering the page on the virtual screen. Based on this representation, they use a number of heuristics for the recognition of the header, footer, menu, and body of the page.

In the paper [50], Liu et al. describe a method to mine data records in a Web page automatically. The algorithm is called MDR (Mining Data Records in Web pages). It finds data records formed by tables and form related tags. A large majority of Web data records are formed by them. The MDR method is based on two observations. The first is that a group of data records that contain descriptions of a set of similar objects are typically presented in a particular region of a page and are formatted using similar HTML tags (sub-tree patterns). The second is that a group of similar data records placed in a specific region is reflected in the tag tree by the fact that they are under one parent node (forest-tree pattern). A Web page may have one data region with data records or a few data regions. Different regions may have different data records. MDR techniques work in three steps: (1) Building a HTML tag tree of the page. (2) Mining data regions in the page using the tag tree and string comparison. (3) Identifying data records from each data region. The experiments showed good results in data region extraction in many domains (books, travel, software, auctions, jobs, electronic products, shopping, and search engine results). There are approaches following MDR (e.g. [69, 73]).

An improvement of the MDR method is presented in [85, 86]. It is called Depta (Data Extraction based on Partial Tree Alignment). Specifically, the Depta method uses visual cues to find data records. Visual information helps the system in two ways: (1) It enables the system to identify gaps that separate data records, which helps to segment data records correctly because the gap within a data record is typically smaller than that in between data records. (2) The proposed system identifies data records by analyzing the HTML DOM tree. The visual information is obtained after the HTML code is rendered by a Web browser; it also contains information about the hierarchical structure of the tags. In this method, rather than analyzing the HTML code, visual information is utilized to infer the structural relationship among tags and to construct a tag tree.

VIPS (VIsion-based Page Segmentation, see Cai et al. [7, 8]) is an algorithm to extract the content structure for a Web page. The algorithm makes full use of page layout features and tries to partition the page at the semantic level. Each node in the extracted tree content structure corresponds to a block of coherent content in the original page. In the VIPS approach, a web page is understood as a finite set of objects or sub-web-pages. All of these objects do not overlap. Each object can be recursively viewed as a sub-web-page and has a subsidiary content structure. A page contains a finite set of visual separators, including horizontal separators and vertical separators. Adjacent visual blocks are separated by these separators. For each visual block, the degree of coherence is defined to measure the extent of its coherence. The vision-based content structure is more likely to provide a semantic partitioning of the page. Every node, especially the leaf node, is more likely to convey a semantic meaning for building a higher semantic via the hierarchy. In the VIPS algorithm, the vision-based content structure of a page is deduced by combining the DOM structure and the visual cues. The segmentation process work on DOM structure and visual information, such as position, background color, font size, font weight, etc. Some approaches follow VIPS, e.g. [80, 13, 91].

Takama and Mitsuhashi present in [71] a visual similarity comparison problem and a visual feature model for web pages. The visual feature model should be as general as possible and based on two principles: (1) Granularity of region segmentation should be adjustable according to applications. (2) Positional relationships among segmented regions should also be considered. The method of Web page layout analysis is to segment a whole page image into several regions, each of which is labeled as either of text, image or a mixture of both. In this method, a page image is divided into several regions based on edge detection. The processing of layout analysis can be summarized as follows: (1) Translates page images from RGB to YCbCr color space. (2) Obtains an edge image from Y image. (3) Obtains an initial region set by connecting neighboring edges. (4) Obtains the second region set by merging small regions in the initial set. (5) Labels each region in it as either text or image. (6) Merges neighboring regions in second region set. If the merged regions have different content types, the resultant region has a label mixture. Experiments on visual similarity comparison using this method showed positive results in the search engine domain and corporate home pages.

A systematic HTML web page segmentation algorithm, specifically designed to segment online journals, is described by Zou et al. in [93]. Based on the observation of online journal articles, the geometric layout is the most important cue for semantic organization. The paper introduces a tree structure model, zone tree, which hierarchically groups the DOM nodes into zones based on the geometric layout of the web page. Visually, a zone is a region on a web page that contains one or more DOM nodes. The nodes are divided and then grouped into zones at different levels depending on their geometric relationships. Similarly to VIPS, the important advantage of the zone tree model is that it is HTML tag independent.

Remarks

1. There is one well known conclusion following the methods of segmentation of Web page. A web page usually contains various contents such as advertisements, navigation, interaction, etc., which are not fully related to the topic of the whole Web page. Furthermore, a web page often contains multiple topics that are not relevant to each other (see [8, 13, 22]). In this case, multi-topic Web pages are divided into several blocks of homogeneous content.
2. This feature of Web pages can be used in different ways. Based on VIPS, in [83], there is a method for how to improve a pseudo-relevant feedback in Web information retrieval.
3. There is a set of approaches about how to find out those Web page parts which are of a higher importance for the user in the semantic sense (e.g. [29, 41]). The procedure is usually the following: the Web page is segmented into individual parts and the quality of the content is evaluated for each segment. In the technical sense it can be noise elimination and selection of text content.

The aforementioned methods have in common is that they analyze the page as a whole and they are primarily aimed at the segmentation of content in a way, to further explore the semantic content of each segment. From this perspective, all these methods can be seen as a technical tool to support the methods listed below. A common feature of the described methods is that the subjects of analysis (detection and extraction) can be named concisely.

There are always one or more sections of a page that is semantically and structurally independent, and often visually separated from other parts of the Web site. An example might be a customer review, a forum, news, product features, product catalog, a table, etc. In addition to these examples, the same logic of usage stands also for other named parts of Web pages, but their analysis is not very significant. Examples are the menu, advertisements, header, footer, etc. Usually, these objects are called noise and there are methods that specialize in the removal of this noise during the analysis process (e.g. Yi et al. [82]).

The following methods are divided into groups according to what is examined and those that occur most frequently are described. But one can find others like tourism and accommodation information (see Kiyavitskaya et al. [39]), book details (see Zhai et al. [85] and Zhang et al. [87]), car technical detils (see [21]), etc.

For the named parts of the page the term MicroGenre is adopted, as will be explained in chapter 4.5.

Genre Detection

The aim of this group of methods is to assign the Web page to some type. This type is either known in advance or it arises as a result of method application. The Genre is the division of concrete forms of art using the criteria relevant to the given form (e.g. film genre, music genre and literature genre). Genres were first introduced to the information system field in the early nineties by Yates and Orlikowski (see [57]).

In document context, the Genre is a taxonomy that incorporates the style; form and content of a document which is orthogonal to topic, with fuzzy classification of multiple genres (see [5]). In the same paper, existing classifications are described. As a result, 115 genres are identified e.g. Ads, Calendars, Classifieds, Comparisons, Contact info, Coupons, Demographics, Documentaries, E-cards, FAQs, Homepages, Hubs, Lists, News, Newsgroups, Reports, etc.

Regarding these classifications there are many approaches to genre identification methods. In paper [65], classification problems are analyzed in terms of two broad textual phenomena: genre hybridism and individualization. The aim of this paper is to show that web pages need a zero to multi-genre classification scheme in addition to the traditional single genre classification.

Kennedy and Shepherd [37] analyzes home page genres (personal homepage, corporate home page or organization home page).

Chaker and Habib [9] propose a flexible approach for Web page genre categorization. Flexibility means that the approach assigns a document to all predefined genres with different weights.

In the paper [36], Kanaris and Stamatatos propose a low-level representation of the style of Web pages based on n-grams. Experiments based on two benchmark genre corpora were presented.

In the paper [61], Rehm analyzes academic Web pages in order to automatically classify them into Web genres. The analysis of a 200 document sample illustrates a notion of Web genre hierarchy into which Web genre types and modules are embedded.

Dong et al. [19] describe a set of experiments to examine the effect of various attributes of web genres on the automatic identification of the genre of web pages.

Four different genres are used in the data set (FAQs, News, E-Shopping and Personal Home Pages).

Rosso [63] explored the use of genre as a document descriptor in order to improve the effectiveness of Web searching. The author of the paper formulates three hypotheses: (1) Users of the system must possess sufficient knowledge of the genre. (2) Searchers must be able to relate the genres to their information needs. (3) The genre must be predictable by a machine applied algorithm because it is not typically explicitly contained in the document.

Kudělka et al. [44] introduce a concept of MicroGenres. Under the notion of MicroGenres, authors understand more or less independent Web page elements, which have some specific purpose and content. Utilizing the term "MicroGenre", the Web page can be described as a collection of MicroGenres. In previous works, authors have focused on Web design pattern detection (see [42]).

Table Extraction

Tables are an important element for structuring related data. The extraction of tables from Web pages appears to be one of the key tasks for further retrieval of structured data. Differentiation of tables emerges as a common problem for all approaches, which represent page layout and those which contain structured data. Tables belong to the field of domain independent objects; however, their extraction can serve well for the detection of other domain dependent objects.

A survey of different approaches is in [84]. Generally, tables often have associated text, including titles, captions, data sources, and footnotes or additional text that elaborate on cells.

Lerman et al. [46] describe an approach to automatic Web table segmentation and extraction of records. This approach is based on the common structure of many Web sites, which present information as lists or tables. Two algorithms are presented that use redundancies in the content of tables. The first one finds the segmentation by solving a constraint satisfaction problem. The second one uses probabilistic inference to find the record segmentation.

In [21], an approach which requires aspects of table understanding is proposed, but it relies especially on extraction ontology. This approach includes five steps: (1) Location of table of interest using values specified in ontology. (2) Setting of attribute-value pairs by individual strings recognized by ontology. (3) Adjustment of attribute-value pairs. (4) Analysis of extraction patterns by the table layout, links to sub-tables, and location of text surrounding the tables. (5) Using the recognized extraction patterns, the system can infer a general mapping from the source to the target.

An approach describing the transformation of arbitrary tables into explicit semantics is presented in [58, 59]. The Tartar system (Transforming Arbitrary TAbles into fRames) is based on a grounded cognitive table model. The presented method is independent of domain knowledge and document types, and enables query answering over heterogeneous tables. HTML tables are analyzed along four aspects: Physical - a description in terms of inter-cell relative location. Structural - the topology of cells as an indicator of their navigational relationship. Functional - the purpose of areas of the tables in terms of data access. Semantic - the meaning of text in the table and the relationship between the interpretation of cell content, the meaning of structures in the table, and the meaning of its reading.

In a paper [26], a different approach is presented . Authors use a tree representation of variation in the two-dimensional visual box model used by web browsers for domain-independent information extraction from web tables. Authors refer to the method as VENTex (Visualized Element Nodes Table extraction). The method works in the following steps: (1) Table location as a task of identifying tables and their constituent logical cells on web pages. (2) Table recognition as a task of identifying the spatial relationships between individual logical cells of a table. (3) Table interpretation as a task of extracting and saving information from tables in a structured format that preserves the meta-information contained in all visual relationships between the individual categories.

Kim and Lee [38] present an efficient method for Web page processing which consists of two phases: area segmentation and structure analysis. The area segmentation cleans up tables and segments them into attribute and value areas by checking visual and semantic coherency. The hierarchical structure between attribute and value areas is then analyzed and transformed into an XML representation using a proposed table model.

Opinion Extraction

There are a wide area of methods which aim to summarize opinions of customers on a product or on its specific features. In individual methods, language analysis (NLP) is also used. Opinions of customers on product Web pages are the main source for analysis, but there can also be discussions on thematic forums or individual reviews in the form of articles (Customer Reviews, Reviews, Discussions, Blogs).

A survey of recent methods of opinion mining is ilustrated in [45]. In this paper, three tasks specific to opinion mining are analyzed: development of linguistic resources, sentiment classification, and opinion summarization.

Another survey is introduced in [14]. The survey is focused on legal blogs (a.k.a. Blawgs) which is any given Weblog that focuses on substantive discussions of the law, the legal profession, including law schools, and the process by which judicial decisions are made. The top-level taxonomy presented shows a variety of topics for the blogging sub-community (General legal blogs, Blogs categorized by legal specialty, Blogs categorized by law or legal event, Blogs categorized by jurisdictional scope, Blogs categorized by author/publisher, Blogs categorized by number of contributors, Miscellaneous blogs categorized by topic, Collections of legal blogs).

In [33], Hu and Liu propose a study of a problem of feature-based opinion summarization of customer reviews of products sold online. The approach has two steps: (1) Identify the features of the product on which customers have expressed opinions and rank the features according to their frequencies. (2) For each feature positive or negative opinions in customer reviews are identified.

A similar approach is described in [51]. Liu et al. focus on online customer reviews and makes two contributions: (1) It proposes a framework for analyzing and comparing consumer opinions of products. (2) A prototype system called Opinion Observer is presented. Experimental results show that the technique is highly effective.

Another similar approach is in [60]. Popescu and Etzioni introduce OPINE, a review-mining system whose components include the use of relaxation labeling to find the semantic orientation of words in the context of given product features. The problem of review mining is decomposed into four main sub-tasks in the Opine system: (1) Identify product features. (2) Identify opinions regarding product features. (3) Determine the polarity of opinions. (4) Rank opinions based on their strength.

Ding et al. [18] propose a holistic lexicon-based approach to solving the problem by exploiting external evidences and linguistic conventions of natural language expressions. This approach allows the handling oh opinion words that are context dependent, which cause major difficulties for existing algorithms.

News Extraction

News extraction methods deal with the extraction of articles from news Web pages. From the method point of view it is a special case of methods from the field of text content extraction. One of the aims of these methods can be to find duplicities published on different Web sites.

Reis et al. [62] present an approach based on the concept of tree-edit distance and allows not only the extraction of relevant text from the pages of a given Web site, but also the fetching of the entire Web site content and the identification of the pages of interest. An algorithm is presented for determining the tree edit distance between trees representing Web pages. In order to extract the news, this approach recognizes common characteristics that are usually present in news portals (a home page that presents some headlines; several sections of pages that provide the headlines divided into areas of interest; and pages that actually present the news, containing the title, author, date and body of the news). The goal of this method is to correctly extract the news, and disregard the other pages.

In paper [88], there is a template-independent news extraction approach that simulates human beings. The approach is based on a stable visual consistency among news pages across Web sites. The presented method is as following: (1) Representation of a DOM node with a block and basic visual features to eliminate the diversity of tags. (2) Extended visual features are calculated as relative visual features to the parent block. (3) Two classifiers are combined for a vision based wrapper for extraction.

Han et al. [31] propose an effective and efficient algorithm to extract the news article contents from the news pages without the analysis of news sites before extraction. They calculate the relevance between the news title and each sentence in the news page to detect the news article contents.

Discussion Extraction

Discussion as an object can be a good source for Opinion extraction. Schuth et al. [68] describe techniques to collect, store, enrich and analyze comments on articles. An extraction method is based on similarity of comments and their recurring features (name, date and time, e-mail, etc.).

A similar approach is in [47]. Limanto et al. present a system based on generated wrappers.

Zhu et al. [92] propose a topic detection and tracking method for the discussion threads. They design a framework, focusing on the very nature of discussion data, including a thread/post activity validation step, a term pos-weighting strategy, and a two-level decision framework considering not only the content similarity but also the user activity information.

Product Details Extraction

Product details are specific objects on product Web pages. It is a basic characteristic, which usually contains a picture, product name, price information, etc. The aim of the methods is to extract as much information as possible. This information can be saved onto a database and then one can implement operations within this database. An extraction of information on product details is a typical task in supervised or semi-supervised approaches.

The goal of Nie et al. [53, 54] is to explore a paradigm to enable web search at the object level, extracting and integrating web information for objects relevant to a specific application domain. Authors introduce the overview of an object-level vertical search engine "Windows Live Product Search". By empirically studying Web pages about the same type of objects across different Web sites, authors have found many template-independent features. Therefore, they proposed template-independent meta-data extraction techniques for the product objects (name, image, price, and description). The purpose of this research is to provide intelligent search results to the user. The same papers introduce an academic search engine Libra (the source can be described as MicroGenre Author and publications) based on the same idea.

A similar approach on extracting author meta-data is proposed in [89]. Zheng et al. attempts to extract author meta-data from their home-pages. They have chosen homepage domains as the source as it is more reliable, comprehensive and more up to date than any other sources. This approach takes advantage of visual features. Based on their observation, visual features are more stable than content-based features for certain types of information which an author puts on the homepage. This is understandable because when people design their homepages, they usually will follow some hidden conventions (in our meaning web usability principles or/and Web design patterns).

Technical Features Extraction

On product or review pages there is usually a list of the technical parameters of the product. Methods of Product technical features mining deal with the extraction of these parameters for selected products and the aim is mainly to be able to compare similar products.

Schmidt et al. [66, 67] present a symbolic approach to extract domain-specific technical features of products from large German corpora. The proposed methods depend on manually added lists of technical measures. The method is a composition of three steps: (1) Extracting product-features for the selected product-class to generate a template for concrete offers. (2) Extracting concrete offers and filling this template, (3) Mapping the selected application to the constraints of technical features and suggesting a set of adequate products. The experiment is presented on a domain of camera technical features.

Wong and Lam describe a similar approach in [79] (experiments on camera and MP3 player features). The approach is a two-phase framework for mining and summarizing hot items in multiple auction Web sites. The objective of the first phase is to extract the product features and product feature values of these items.

4.5   MicroGenre

In all sectors of the arts, the Genres are vague categories without fixed boundaries and are especially formed by sets of conventions. Many artworks are cross-genre and employ and combine these conventions. Probably the most deeply theoretically studied Genres are the literary ones. It allows us to systematize the world of literature and consider it as a subject of scientific examination. One can find the term MicroGenre in this field. For example, in [52] the MicroGenre is seen as part of a combined text. This term has been introduced to identify the contribution of inserted Genres to the overall organization of text. The motivation of using the term "MicroGenre" is because it is used as a building block for the analytic descriptive system. On the Web, MicroGenre can be used more technically and can provide a means for an effective solution for Web pages. From our point of view the Web page is structured similarly to a literacy text using parts which are relatively independent and have their own purpose (see fig. 4.3). For these parts the term "MicroGenre" was chosen.

Definition 4.2. (Web) MicroGenre is a part of a Web page,

1. whose purpose is general and repeats itself frequently
2. which can be named intelligibly and more or less unambiguously so that the name is understandable for the Web page user (developer, designer, etc.)
3. which is detectable on a Web page using a computer algorithm

The MicroGenre can, but does not have to, strictly relate to the structure of a Web page in a technical sense. E.g. it does not necessarily have to be represented by one subtree in the DOM tree of the page or by one block in the sense of the visual layout

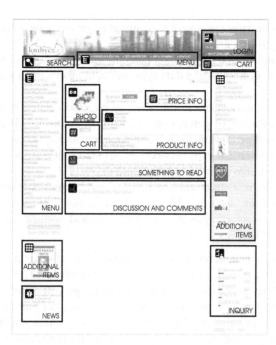

Fig. 4.3 Structure of Web page. Contemporary Web pages are often very complex, neverthe-less they can usually be described using several MicroGenres. This kind of description can be more flexible than the Genre description (which usually represents the whole page)

of the page. Instead it can be represented by one or more segments of a page, which unite it.

Remark

MicroGenres are also contexts which encapsulate related information. In paper [43] we show the way in which we extract snippets from individual MicroGenres. We use these snippets in our Web application as additional information for Web page description. The detection of MicroGenres can be considered in a similar way as to a first step in using Web information extraction methods (see [11]).

4.5.1   Pattrio Method

It follows the previous description that in order to be able to speak about the Mi-croGenre, this element has to be distinguishable to the user. From what attributes should the user recognize, if and what MicroGenre there is or is in question? We work with up to three levels of view:

1. The first view is purely semantic in the sense of the textual content of a page. It does not always have to be a meaning in a sense of natural language such

as sentences or paragraphs with a meaningful content. Logically coherent data blocks can still lack grammar (see [91]).

For example, Price information can only be a group of words and symbols ('price', 'vat', symbol $) of a data-type (price, number). For a similar approach see [64].

2. The second view is visual in the sense of page perception as a whole. Here individual segments of perception or groups of segments of the page are in question. It is dependent on the use of colors, font and auxiliary elements (lines, horizontal and vertical gaps between the segments etc.) Approaches based on visual analysis of Web pages can be found in [7], [71].

3. The third view is a structural one in a technical sense. It is about the use of special structures, such as tables, links, navigation trees, etc. There are approaches based on analysis of the DOM tree and special structures such as tables [50], [59].

The first view is dependent on the user's understanding of the text stated on a Web page. The second and third views are independent of this user ability. However, it can be expected that an Arabic or Chinese product page will be recognized also by an English-speaking user who does not have a command of those languages. It is determined by the fact that for the implementation of certain intentions there are habitual procedures which provide very similar results regardless of the language. On the other hand, if the user understands the page, he/she can focus more on the semantic content of the MicroGenre. For example, in the case of Product main info, the user can read what the product is in question, its price and on what conditions it can be purchased.

Remark

In our previous work, we have focused on Web design pattern detection. The research findings illustrated that it is useful to generalize the detection algorithm on all objects that provide the benefits to the users. These objects can be out of the scope to what is useful to the Web developers, but can be helpful in a similar way to that of Genres. Therefore we have chosen the term "MicroGenre" for our further research.

4.5.2 Analysis

In our approach, there are elements with semantic contents (words or simple phrases and data types) and elements with importance for the structure of the web page where the MicroGenre instance can be found (technical elements). The rules are in the way that individual elements take part in the MicroGenre display. While defining these rules, we were inspired by the Gestalt principles. We formulated four rules based on these principles. The first (proximity) defines the acceptable measurable distances of individual elements from each other. The second (closure) defines the way of creating independent closed segments containing elements. One or more segments then create the MicroGenre instance on the web page. The third (similarity) implies suggests that the MicroGenre includes more related segments that are similar. The fourth (continuity) establishes that the MicroGenre contains more various segments that together create the Web pattern instance. The relations among MicroGenres can be on various levels similar to the patterns in pattern languages (especially association and aggregation).

Remark

Other objects in the Web page analysis can be used. E.g. Dujovne and Velásquez [20] introduce a methodology for discovering Web page key objects based in both Web usage and content mining. Key objects could be any text, image or video present in a web page, that are the most appealing objects to users.

Lightweight Formalization

For the purpose of the proposal of the detection algorithm, a lightweight formalization was used. This formalization is a simple one and it is only a starting point for the description of a MicroGenre.

Let us have a MicroGenre MG. Then let us mark the set of MicroGenres words by W, the set of data types by D and the set of technical (HTML) elements by T. Let us mark the set of MicroGenre rules by Ru. Let us mark the set of relations of the MicroGenre MG to other MicroGenres by Re. So, the MicroGenre MG is defined as $MG = (W, D, T, Ru, Re)$. It is given that x is the Web page containing the instance of this MicroGenre. Then the instance $I(x)$ of the MicroGenre MG on the Web page x is

$$I(x) = (W(x), D(x), T(x), Ru(x), Re(x)),$$

where $W(x) \subset W$, $D(x) \subset D$, $T(x) \subset T$, $Ru(x) \subset Ru$, $Re(x) \subset Re$, while these elements fulfill the rules Ru of the MicroGenre N.

Example (Discussion)

This formalization can be seen in a simplified example.

Words: {Main: re, reply, discussion, forum, author, question, answer, thread contribution,, subject, sent; Complementary: date, name, post, topic}
Data types: {Main: date, time, first name; Complementary:}
Technical elements: {link, label, input}
Rules: {proximity: 16; closure: normal; similarity: high; continuity: low}
Associations: {Contains: Short Paragraphs; Uses: Date per Paragraph; Complements: Review and Comments}

The basic algorithm for the detection of MicroGenres then implements the preprocessing of the code of the HTML page (only selected elements are preserved – e.g. block elements such as tables, div, lines, etc., see Table 4.2), segmentation and evaluation of rules and associations. The result for the page is the score of

Table 4.2 HTML tags - classification for analysis

Types	Tags
Headings	H1, H2, H3, H4, H5, H6
Text containers	P, PRE, BLOCKQUOTE, ADDRESS
Lists	UL, OL, LI, DL, DIR, MENU
Blocks	DIV, CENTER, FORM, HR, TABLE, BR
Tables	TR, TD, TH, DD, DT
Markups	A, IMG
Forms	LABEL, INPUT, OPTION

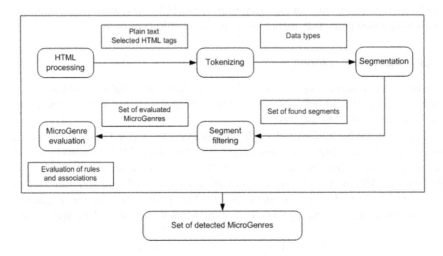

Fig. 4.4 MicroGenre detection process

MicroGenres that is present on the page. The score then says what the probability is of expecting the MicroGenre instance on the page for the user. The entire process, including MicroGenre detection, is displayed in Figure 4.4.

4.6 Experiments

10 MicroGenres was used for our experiments in this chapter (Price information, Purchase possibility, Special offers, Hire sale, Technical features, Discussion, Review, Second hand, Login, and Inquiry). The names of MicroGenres emerged from the discussions between us and students that took part in our experiments. For another experiment, especially in our research concerning social aspects of Web page contents, more than 20 MicroGenres (see [44]) were prepared.

4.6.1   Accuracy of Pattrio Method

The aim of the experiment was to find out what information could be found using this method on the web pages returned from the Google search engine. The Pattrio method evaluated the relevance of each returned web page with respect to the query and the user's expectation. For this experiment three products were selected (we typed the product id and keywords into the search engine) - *Apple iPod Nano 1GB, Canon EOS 20D, Star Wars Trilogy film.*

The emphasis was placed on querying products that are common and are sold in high quantities. One of the most requested queries is that of product information which discusses the product and issues around purchasing it. Therefore our experiment focused on two MicroGenres, the "Discussion" and the "Purchasing possibility". For each MicroGenre a simple query was guilt that consisted of the product id and a keyword for the MicroGenre (Discussion and Purchase possibility). Six groups of pages were provided after querying Google. From each result set only the first 100 pages were used. As users pages with discussions and purchasing possibilities were expected. First the pages for experiments were manually checked and those that did not fit our requirements were removed - pages with broken links and documents, were not evaluated (pdf, doc, xml). The remainder of the pages were evaluated using a three-degree scale:

+   Page contains required object.
-   Page does not contain required object.
?   Unable to evaluate results.

The manual evaluation is not always precise. The reason is that these MicroGenres do not consist of a precise formalization and are only a description of a common experience of the user and the web designer. Therefore each page was evaluated with a group of three students. Each student had to classify each page. If one of them had a different mark for a page than the others the document were classified as *unable to evaluate the results* (?). The results are shown in Figure 4.5.

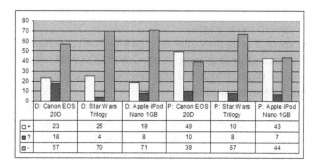

	D: Canon EOS 20D	D: Star Wars Trilogy	D: Apple iPod Nano 1GB	P: Canon EOS 20D	P: Star Wars Trilogy	P: Apple iPod Nano 1GB
□ +	23	25	19	49	10	43
■ ?	18	4	8	10	8	7
▣ -	57	70	71	39	67	44

Fig. 4.5 Manually evaluated pages

Then the sets of pages were processed again but through our method, which computes the score of the MicroGenre presence on the page. Let the score (thresholds are set up according to the results of our experiments) of MicroGenre be V then:

- if $V \geq \frac{2}{3}$ then MicroGenre is found (+)
- if $V \leq \frac{1}{4}$ then MicroGenre is not found (-)
- if $\frac{1}{4} \leq V \leq \frac{2}{3}$ then it is not possible to decide (?)

The results are shown in Figure 4.6. There are values in each column, whose manual evaluation corresponds to the evaluation made by the Pattrio method.

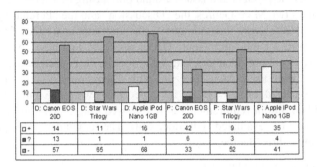

	D: Canon EOS 20D	D: Star Wars Trilogy	D: Apple iPod Nano 1GB	P: Canon EOS 20D	P: Star Wars Trilogy	P: Apple iPod Nano 1GB
□ +	14	11	16	42	9	35
■ ?	13	1	1	6	3	4
□ -	57	65	68	33	52	41

Fig. 4.6 Evaluation by Pattrio method

Experiment Evaluation

For the evaluation of the accuracy of our method a formula was used which works with the count of manual and automatically evaluated pages. The formula for computation of *retrieval accuracy* (see [70]) was also used.

$$RA = \frac{relevant\ pages\ retrieved\ in\ top\ T\ returns}{T}$$

where T is the count of found pages.

In this experiment T was the count of pages in one set (returned by a search engine) with the same value (e.g. pages with Canon EOS 20D with discussion and value +). Only those pages which had the same value for the manual and for the automatic evaluation were considered relevant.

Let S_i^v be the i-th set of pages manually evaluated with value v, where $v \in \{+, ?, -\}$. Let SM_i^v be a subset of S_i^v for which the manual evaluation and the evaluation using the Pattrio method have the same value. The overall method accuracy for one particular set of pages can be computed according to the formula:

$$m(S_i^v) = m_i^v = \frac{|SM_i^v|}{|S_i^v|}$$

In Figure 4.7 the method accuracy is in percentage form for each set of pages evaluated in our experiment (we worked with 6 groups of pages and each with 3 evaluated sets).

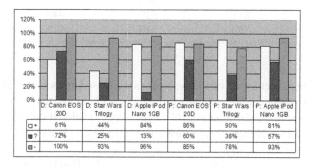

Fig. 4.7 Method accuracy in percentage

For example the first value 61% expresses the method accuracy for the pages with Canon EOS 20D product in which there was a discussion. The value was computed according to the formula

$$m(S_1^+) = m_1^+ = \frac{|SM_1^+|}{|S_1^+|} = \frac{14}{23} \doteq 0.609$$

where values 14 and 23 are from tables in Figures 4.5 and 4.6. The method accuracy is between 13 and 100 percent.

For the overall evaluation of method accuracy in our experiment formula computing was used with counts of manual evaluated pages for each set. The method accuracy for the selected sets of pages is computed according to the formula:

$$M = \frac{\sum\limits_{[i,v]} |S_i^v| \, m_i^v}{\sum\limits_{[i,v]} |S_i^v|}$$

where m_i^v is the evaluation of one concrete set. In Table 4.3 there are result values for the pages with discussions, purchase possibility and overall method accuracy noted in percentages.

For example, the value 96% expresses the method accuracy in the situation when the web page does not consist of a discussion. The value is computed according to the formula:

$$M = \frac{n_1^- m_1^- + n_2^- m_2^- + n_3^- m_3^-}{n_1^- + n_2^- + n_3^-} \doteq \frac{57 \cdot 1 + 70 \cdot 0.93 + 71 \cdot 0.96}{57 + 70 + 71} \doteq 0.96$$

where each value is taken from the tables in Figures 4.5 and 4.7.

Table 4.3 Method accuracy

Microgenre	+	?	−	Total
Discussion	61%	50%	96%	83%
Purchase	84%	52%	84%	81%
Total	75%	51%	91%	82%

For this experiment a total of 572 web pages were used and which were manually computed and evaluated. The manual evaluation was compared to the Pattrio method evaluation. In the manual evaluation 169 pages were considered suitable (containing MicroGenre), 348 were unsuitable (not containing MicroGenre) and 55 of which were deemed hard to classify.

The method accuracy for Discussion is 83% and for Purchase possibility 81%. Pages containing objects the method accuracy scored 75% and for pages that do not contain objects 84%. Regarding the situations deemed hard to decide the method accuracy is 52% and the overall method accuracy 82%.

4.6.2 Analysis by Nonnegative Matrix Factorization

The nonnegative matrix factorization (NMF) method for text mining is a technique for clustering that identifies semantic features in a document collection and groups the documents into clusters on the basis of shared semantic features [17]. A collection of documents can be represented as a term-by-document matrix. Since each vector component is given a positive value if the corresponding term is present in the document and a zero value otherwise, the resulting term-by-document matrix is always nonnegative. This data non-negativity is preserved by the NMF method as a result of constraints that produce nonnegative lower rank factors that can be interpreted as semantic features or patterns in the text collection.

NMF Method

With the standard vector space model a set of documents S can be expressed as an $m \times n$ matrix V, where m is the number of terms and n is the number of documents in S. Each column V_j of V is an encoding of a document in S and each entry v_{ij} of vector V_j is the value of the i-th term with regard to the semantics of V_j, where i ranges across the terms in the dictionary. The NMF problem is defined as finding an approximation of V in terms of some metric (e.g., the norm) by factoring V into the product WH of two reduced-dimensional matrices W and H [4]. Each column of W is a basis vector. It contains an encoding of a semantic space or concept from V and each column of H contains an encoding of the linear combination of the basis vectors that approximates the corresponding column of V. Dimensions of W and H are $m \times k$ and $k \times n$, where k is the reduced rank. Usually k is chosen to be much smaller than n. Finding the appropriate value of k depends on the application and is also influenced by the nature of the collection itself.

Common approaches to NMF obtain an approximation of V by computing a (W,H) pair to minimize the Frobenius norm of the difference $V - WH$. The matrices W and H are not unique. Usually H is initialized to zero and W to a randomly generated matrix where each $W_{ij} > 0$. These initial values are improved with iterations of the algorithm.

GD-CLS Method

GD-CLS is a hybrid method that combines some of the better features of other methods. The multiplicative method, which is basically a version of the gradient descent optimization scheme, is used at each iterative step to approximate the basis vector matrix W. H, which is calculated using a constrained least squares (constrained least squares - CLS) model as the metric.

Algorithm

1. Initialize W and H with nonnegative values, and scale the columns of W to unit norm.
2. Iterate until convergence or after l iterations:

   - $W_{ic} = W_{ic} \frac{(VH^T)_{ic}}{(WHH^T)_{ic}+\varepsilon}$, for c and i $[\varepsilon = 10^{-9}]$
   - Rescale the columns of W to unit norm
   - Solve the constrained least squares problem where $min_{H_j}\{||V_j - WH_j||_2^2 + \lambda||H_j||_2^2$ the subscript j denotes the j-th column, for $j = 1,\ldots,m$. Any negative values in H_j are set to zero. The parameter k is a regularization value that is used to balance the reduction of the metric $||V_j - WH_j||_2^2$ with the enforcement of smoothness and sparsity in H.

For any given matrix V, matrix W has k columns or basis vectors that represent k clusters, matrix H has n columns that represent n documents. A column vector in H has k components, each of which denotes the contribution of the corresponding basis vector to that column or document. The clustering of documents is then performed based on the index of the highest value of k for each document. For document i $(i = 1,\ldots,n)$, if the maximum value is the j-th entry $(j = 1,\ldots,k)$, document i is assigned to cluster j. We used the GD-CLS method for searching $k = 2,3,4,5$ clusters. Results are in Table 4.4.

One can see that in the table the results are readable. For example in the first row is a vector which represents sale - price information, purchase possibility, special offer, hire sale, technical features, and login. In the second row is a vector which describes information cluster - review, discussion, login, and inquiry. We set the limit for a successful MicroGenre to 0.05. The method was unstable for 6 or more clusters. In the table there are clusters-vectors with only one higher value (for example row 5). These clusters do not have a positive information value because one expects at least two MicroGenre on each page.

Table 4.4 Clustering by NMF

clusters	price info	purchase	special offer	hire sale	features	review	discussion	login	second hand	inquiry
1 / 2	0	0.01	0	0.01	0.03	0.29	0.27	0.337	0	0.06
2 / 2	0.33	0.25	0.19	0.06	0.07	0	0	0.05	0.03	0.01
1 / 3	0.35	0.26	0.21	0.07	0.07	0	0	0	0.04	0.01
2 / 3	0.02	0	0	0.01	0.01	0.41	0.46	0	0.02	0.06
3 / 3	0	0.08	0	0	0.08	0.04	0	0.76	0	0.04
1 / 4	0.42	0	0.41	0.04	0	0	0	0	0.13	0.01
2 / 4	0.01	0	0	0.02	0.02	0.42	0.47	0	0.01	0.06
3 / 4	0	0	0.01	0	0.05	0.04	0	0.85	0	0.05
4 / 4	0.26	0.49	0	0.09	0.15	0.01	0	0	0	0
1 / 5	0	0	0	0.02	0.02	0.42	0.47	0	0	0.06
2 / 5	0.25	0.50	0	0.09	0.16	0	0	0	0	0
3 / 5	0.31	0.01	0.57	0.06	0	0	0	0	0.04	0.01
4 / 5	0.62	0	0	0	0	0	0	0	0.36	0.02
5 / 5	0.01	0	0	0	0.04	0.04	0	0.85	0	0.05

4.7 Summary

In this chapter were discussed the development of methods from the field of Web content mining, especially regarding their possible use for the search queries and comprehensible description of the Web pages. These methods have this in common: each analyzed part of the Web page (object) can be named concisely. The term MicroGenre was adopted for these objects. From an abstract point of view MicroGenres can be seen as information granules, which simplify the appearence of Web pages. Using this approach different methods can be applied from the field of granular computing. Our experiments show that MicroGenres can be detected successfully and in a relatively simple manner. After that, the page can be represented in such a form that the standard mathematical methods can be applied. This has been highlighted in the experiment applying a variant of the NMF method on an extensive set of Web pages from a selling domain.

Our experiments have also shown that one of the qualities of MicroGenres is their small language and cultural dependence. Our current research leads to the usage of MicroGenres for cross-language searching.

References

1. Alexander, C.: A Pattern Language: Towns, Buildings, Construction. Oxford University Press, New York (1977)
2. Alexander, C.: The Timeless Way of Building. Oxford University Press, Oxford (1979)
3. Baumgartner, R., Flesca, S., Gottlob, G.: Visual Web Information Extraction with Lixto. In: Proc. of the 27th Int. Conference on Very Large Data Bases, pp. 119–128 (2001)
4. Shahnaz, F., Berry, M.W., Pauca, P.V., Plemmons, R.J.: Document clustering using non-negative matrix factorization. Information Processing and Management 42(2), 373–386 (2006)

5. Boese, E.S., Howe, A.E.: Effects of web document evolution on genre classification. In: Proceedings of the 14th ACM international Conference on information and Knowledge Management, CIKM 2005, Bremen, Germany, October 31 - November 05, 2005, pp. 632–639. ACM, New York (2005)

6. Borchers, J.O.: A pattern approach to interaction design. AI & Society 15(4), 359–376 (2001)

7. Cai, D., Yu, S., Wen, J.-R., Ma, W.-Y.: Extracting Content Structure for Web Pages based on Visual Representation. In: Zhou, X., Zhang, Y., Orlowska, M.E. (eds.) APWeb 2003. LNCS, vol. 2642, pp. 406–417. Springer, Heidelberg (2003)

8. Cai, D., Yu, S., Wen, J.-R., Ma, W.-Y.: VIPS: a visionbased page segmentation algorithm, Microsoft Technical Re-port, MSR-TR-2003-79 (2003)

9. Chaker, J., Ounelli, H.: Genre Categorization of Web Pages. In: ICDM Workshops 2007, pp. 455–464 (2007)

10. Chang, C., Lui, S.: IEPAD: information extraction based on pattern discovery. In: Proceedings of the 10th international Conference on World Wide Web, WWW 2001, Hong Kong, May 01-05, 2001, pp. 681–688. ACM, New York (2001)

11. Chang, C.H., Kayed, M., Girgis, M.R., Shaalan, K.F.: A Survey of Web Information Extraction Systems. IEEE Transactions on Knowledge and Data Engineering 18(10), 1411–1428 (2006)

12. Chen, J., Zhou, B., Shi, J., Zhang, H., Fengwu, Q.: Function-based object model towards Website adaptation. In: Proceedings of the 10th international conference on World Wide Web, Hong Kong, pp. 587–596 (2001)

13. Chibane, I., Doan, B.L.: A web page topic segmentation algorithm based on visual criteria and content layout. In: SIGIR 2007, pp. 817–818 (2007)

14. Conrad, J.G., Schilder, F.: Opinion mining in legal blogs. In: Proceedings of the 11th international Conference on Artificial intelligence and Law, ICAIL 2007, Stanford, California, June 04-08, 2007, pp. 231–236. ACM, New York (2007)

15. Cosulschi, M., Constantinescu, N., Gabroveanu, M.: Classifcation and comparison of information structures from a web page. The Annals of the University of Craiova 31, 109–121 (2004)

16. Dearden, A., Finlay, J.: Pattern Languages in HCI: A critical review. Human Computer Interaction 21(1), 49–102 (2006)

17. Ding, C., Li, T., Peng, W., Park, H.: Orthogonal nonnegative matrix t-factorizations for clustering. In: ACM SIGKDD International Conference on Knowledge Discovery and Data Mining (KDD 2006), pp. 126–135. ACM Press, New York (2006)

18. Ding, X., Liu, B., Yu, P.S.: A Holistic Lexicon-Based Approach to Opinion Mining. In: Web Search and Web Data Mining, Palo Alto, California, USA, pp. 231–240 (2008)

19. Dong, L., Watters, C.R., Duffy, J., Shepherd, M.: An Examination of Genre Attributes for Web Page Classification. In: HICSS 2008, p. 133 (2008)

20. Dujovne, L.E., Velásquez, J.D.: Design and Implementation of a Methodology for Identifying Website Keyobjects. In: Velásquez, J.D., Ríos, S.A., Howlett, R.J., Jain, L.C. (eds.) KES 2009. LNCS, vol. 5711, pp. 301–308. Springer, Heidelberg (2009)

21. Embley, D.E., Tao, C., Liddle, S.W.: Automating the extraction of data from HTML tables with unknown structure. Data Knowl. Eng. 54(1), 3–28 (2005)

22. Fernandes, D., de Moura, E.S., Ribeiro-Neto, B.: Computing Block Importance for Searching on Web Sites. In: ACM International Conference on Information and Knowledge Management, Lisboa, Portugal, pp. 165–173 (2007)

23. Flieder, K., Mödritscher, F.: Foundations of a pattern language based on Gestalt principles. In: CHI 2006 Extended Abstracts on Human Factors in Computing Systems, Montreal, Quebec, Canada, April 22 - 27, pp. 773–778. ACM, New York (2006)

24. Gagneux, A., Eglin, V., Emptoz, H.: Quality Approach of Web Documents by an Evaluation of Structure Relevance. In: Proceedings of WDA 2001, pp. 11–14 (2001)
25. Gamma, E., Helm, R., Johnson, R., Vlissides, J.: Design Patterns Elements of Reusable Object-Oriented Software. Addison-Wesley, Reading (1995)
26. Gatterbauer, W., Bohunsky, P., Herzog, M., Krüpl, B., Pollak, B.: Towards domain-independent information extraction from web tables. In: Proceedings of the 16th international Conference on World Wide Web, WWW 2007, Banff, Alberta, Canada, May 08–12, pp. 71–80. ACM, New York (2007)
27. Goldberg, J.H., Stimson, M.J., Lewenstein, M., Scott, N., Wichansky, A.M.: Eye tracking in web search tasks: design implications. In: Proceedings of the 2002 Symposium on Eye Tracking Research & Applications, ETRA 2002, New Orleans, Louisiana, March 25-27, pp. 51–58. ACM, New York (2002)
28. Graham, L.: A pattern language for web usability. Addison-Wesley, Reading (2003)
29. Gupta, S., Kaiser, G., Neistadt, D.,, Grimm, P.: DOM-based Content Extraction of HTML Documents. In: World Wide Web conference (WWW 2003), Budapest, Hungary, pp. 207–214 (2003)
30. Han, J., Kamber, M.: Data mining: concepts and techniques. Morgan Kaufmann Publishers Inc., San Francisco (2000)
31. Han, H., Noro, T., Tokuda, T.: An Automatic Web News Article Contents Extraction System Based on RSS Feeds. Journal of Web Engineering 8(3), 268–284 (2009)
32. Han, J., Chang, K.: Data Mining for Web Intelligence. Computer 35(11), 64–70 (2002)
33. Hu, M., Liu, B.: Mining and summarizing customer reviews. In: Proceedings of the Tenth ACM SIGKDD international Conference on Knowledge Discovery and Data Mining, KDD 2004, Seattle, WA, USA, August 22-25, pp. 168–177. ACM, New York (2004)
34. Ivory, M.Y., Megraw, R.: Evolution of Web Site Design Patterns. ACM Transactions on Information Systems 23(4), 463–497 (2005)
35. Ivory, M.Y., Sinha, R.R., Hearst, M.A.: Empirically validated web page design metrics. In: Proceedings of the SIGCHI conference on Human factors in computing systems, Seattle, Washington, United States, March 2001, pp. 53–60 (2001)
36. Kanaris, I., Stamatatos, E.: Webpage Genre Identification Using Variable-Length Character n-Grams Tools with Artificial Intelligence, 2007. In: ICTAI 2007, pp. 3–10 (2007)
37. Kennedy, A., Shepherd, M.: Automatic identification of home pages on the web. In: Annual Hawaii International Conference on System Sciences (HICSS 2005), pp. 236–251 (2005)
38. Kim, Y.S., Lee, K.H.: Extracting logical structures from HTML tables. Computer Standards & Interfaces 30(5), 296–308 (2008)
39. Kiyavitskaya, N., Zeni, N., Mich, L., Cordy, J.R., Mylopoulos, J.: Text mining through semi automatic semantic annotation. In: Reimer, U., Karagiannis, D. (eds.) PAKM 2006. LNCS (LNAI), vol. 4333, pp. 143–154. Springer, Heidelberg (2006)
40. Kosala, K., Blockeel, H.: Web Mining Research: A Survey. SIGKDD Explorations 2(1), 1–15 (2000)
41. Kovacevic, M., Diligenti, M., Gori, M., Milutinovic, V.: Recognition of Common Areas in a Web Page Using Visual Information: a possible application in a page classification. In: Proceedings of the 2002 IEEE International Conference on Data Mining (ICDM 2002), p. 250 (2002)
42. Kudělka, M., Snášel, V., Lehečka, O., El-Qawasmeh, E.: Semantic Analysis of Web Pages Using Web Patterns. In: Web Intelligence 2006, pp. 329–333 (2006)

43. Kudělka, M., Snášel, V., Lehečka, O., El-Qawasmeh, E., Pokorný, J.: Web Pages Re-ordering and Clustering Based on Web Patterns. In: Geffert, V., Karhumäki, J., Bertoni, A., Preneel, B., Návrat, P., Bieliková, M. (eds.) SOFSEM 2008. LNCS, vol. 4910, pp. 731–742. Springer, Heidelberg (2008)
44. Kudelka, M., Snasel, V., Horak, Z., Abraham, A.: Social Aspects of Web Page Contents. In: IEEE CASoN 2009, Fontainebleau, France, pp. 80–87 (2009)
45. Lee, D., Jeong, O., Lee, S.: Opinion mining of customer feedback data on the web. In: Proceedings of the 2nd international Conference on Ubiquitous information Manage-ment and Communication, ICUIMC 2008, Suwon, Korea, January 31 - February 01, pp. 230–235. ACM, New York (2008)
46. Lerman, K., Getoor, L., Minton, S., Knoblock, C.: Using the structure of Web sites for automatic segmentation of tables. In: Proceedings of the 2004 ACM SIGMOD interna-tional Conference on Management of Data, SIGMOD 2004, Paris, France, June 13-18, pp. 119–130. ACM, New York (2004)
47. Limanto, H.Y., Giang, N.N., Trung, V.T., Zhang, J., He, Q., Huy, N.Q.: An information extraction engine for web discussion forums. In: Special interest Tracks and Posters of the 14th international Conference on World Wide Web, WWW 2005, Chiba, Japan, May 10-14, pp. 978–979. ACM, New York (2005)
48. Liu, B.: Web content mining (tutorial). In: Proceedings of the 14th International Confer-ence on World Wide Web (2005)
49. Liu, B., Chang, K.C.-C.: Editorial: Special Issue on Web Content Mining. IGKDD Ex-plor. Newsl. 6(2), 1–4 (2004)
50. Liu, B., Grossman, R., Zhai, Y.: Mining data records in Web pages. In: KDD 2003, pp. 601–606 (2003)
51. Liu, B., Hu, M., Cheng, J.: Opinion observer: analyzing and comparing opinions on the Web. In: Proceedings of the 14th international Conference on World Wide Web, WWW 2005, Chiba, Japan, May 10-14, pp. 342–351. ACM, New York (2005)
52. Martin, J.R.: Text and clause: Fractal resonance. Text 15, 5–42 (1995)
53. Nie, Z., Wen, J.-R., Ma, W.-Y.: Object-level Vertical Search. In: CIDR 2007, Asilomar, CA, pp. 235–246 (2007)
54. Nie, Z., Ma, Y., Shi, S., xWen, J.-R., Ma, W.-Y.: Web Object Retrieval. In: WWW 2007, pp. 81–90 (2007)
55. Nielsen, J.: DesigningWeb Usability: The Practice of Simplicity. New Riders Publisher, Indianapolis (2000)
56. Nielsen, J., Loranger, H.: Prioritizing Web Usability. New Riders Press, Berkeley (2006)
57. Yates, J., Orlikowski, W.J.: Genres of Organizational Communication: A Structura-tional Approach to Studying Communication and Media. Academy of Management Re-view 17(2), 299–326 (1992)
58. Pivk, A., Cimiano, P., Sure, Y.: From tables to frames. Journal of Web Semantics 3(2-3), 132–146 (2005)
59. Pivk, A., Cimiano, P., Sure, Y., Gams, M., Rajkovic, V., Studer, R.: Transforming arbi-trary tables into logical form with TARTAR. Data Knowl. Eng. 60(3), 567–595 (2007)
60. Popescu, A., Etzioni, O.: Extracting product features and opinions from reviews. In: Proceedings of the Conference on Human Language Technology and Empirical Methods in Natural Language Processing, Vancouver, British Columbia, Canada, October 06 - 08, pp. 339–346. Association for Computational Linguistics, Morristown (2005)
61. Rehm, G.: Towards Automatic Web Genre Identification. In: 35th Annual Hawaii Inter-national Conference on System Sciences (HICSS 2002), vol. 4, p. 101 (2002)
62. Reis, D.C., Golgher, P.B., Silva, A.S., Laender, A.F.: Automatic web news extraction us-ing tree edit distance. In: WWW 2004: Proceedings of the 13th international conference on World Wide Web, pp. 502–511. ACM Press, New York (2004)

63. Rosso, M.A.: User-based identification of Web genres. JASIST (JASIS) 59(7), 1053–1072 (2008/2009)
64. Santini, M.: Description of 3 feature sets for automatic identification of genres in web pages (2006),
    http://www.nltg.brighton.ac.uk/home/Marina.Santini/
    three_feature_sets.pdf
    (last accessed February 02, 2010)
65. Santini, M.: Characterizing Genres of Web Pages: Genre Hybridism and Individualization. In: HICSS 2007, p. 71 (2007)
66. Schmidt, S., Stoyan, H.: Web-based Extraction of Technical Features of Products. GI Jahrestagung (1), 246–250 (2005)
67. Schmidt, S., Mandl, S., Ludwig, B., Stoyan, H.: Product-advisory on the web: An information extraction approach. In: Artificial Intelligence and Applications 2007, pp. 678–683 (2007)
68. Schuth, A., Marx, M., de Rijke, M.: Extracting the discussion structure in comments on news-articles. In: Proceedings of the 9th Annual ACM international Workshop on Web information and Data Management, WIDM 2007, Lisbon, Portugal, November 09-09, pp. 97–104. ACM, New York (2007)
69. Simon, K., Lausen, G.: ViPER: augmenting automatic information extraction with visual perceptions. In: Proceedings of the 14th ACM international Conference on information and Knowledge Management, CIKM 2005, Bremen, Germany, October 31 - November 05, pp. 381–388. ACM, New York (2005)
70. Su, Z., Zhang, H.J., Li, S., Ma, S.: Relevance feedback in content-based image retrieval: Bayesian framework, feature subspaces, and progressive learning. IEEE Transactions on Image Processing 12(8), 924–937 (2003)
71. Takama, Y., Mitsuhashi, N.: Visual Similarity Comparison for Web Page Retrieval. In: Web Intelligence, pp. 301–304 (2005)
72. Tidwell, J.: Designing Interfaces: Patterns for Effective Interaction Design. O'Reilly Media, Inc., Sebastopol (2006)
73. Tseng, Y.-F., Kao, H.-K.: The Mining and Extraction of Primary Informative Blocks and Data Objects from Systematic Web Pages. In: 2006 IEEE/WIC/ACM International Conference on Web Intelligence (WI 2006 Main Conference Proceedings) (WI 2006), pp. 370–373 (2006)
74. Van Duyne, D.K., Landay, J.A., Hong, J.I.: The Design of Sites: Patterns, Principles, and Processes for Crafting a Customer-Centered Web Experience. Pearson Education, London (2002)
75. Vredenburg, K., Isensee, S., Righi, C.: User-Centered Design: An Integrated Approach. Prentice Hall, Upper Saddle River (2002)
76. W3C Document Object Model, http://www.welie.com (last accessed February 02, 2010)
77. Van Welie, M., van der Veer, G.: Pattern Languages in Interaction Design: Structure and Organization. In: Rauterberg, Menozzi, Wesson (eds.) Proceedings of Interact 2003, Zürich, Switzerland, September 1–5, pp. 527–534. IOS Press, Amsterdam (2003)
78. Van Welie, M.: Pattern in Interaction Design, http://www.welie.com (last accessed February 28, 2010)
79. Wong, T.-L.W., Lam, W.: Hot Item Mining and Summarization from Multiple Auction Web Sites. In: ICDM 2005, New Orleans, Louisiana, USA, pp. 797–800 (2005)
80. Xiang, P., Yang, X., Shi, Y.: Effective Page Segmentation Combining Pattern Analysis and Visual Separators for Browsing on Small Screens. In: Web Intelligence 2006, pp. 831–840 (2006)

81. Yang, Y., Chen, Y., Zhang, H.J.: HTML Page Analysis Based on Visual Cues. In: International Conference on Document Analysis and Recognition, 2001, pp. 859–864 (2003)
82. Yi, L., Liu, B., Li, X.: Eliminating noisy information in Web pages for data mining. In: International Conference on Knowledge Discovery and Data Mining, KDD 2003, Washington, DC, USA, pp. 296–305 (2003)
83. Yu, S., Cai, D., Wen, J.-R., Ma, W.-Y.: Improving Pseudo-Relevance Feedback in Web Information retrieval Using Web Page Segmentation. In: The Proceedings of Twelfth World Wide Web conference (WWW 2003), Hungary, pp. 203–211 (2003)
84. Zanibbi, R., Blostein, D., Cordy, J.R.: A survey of table recognition: Models, observations, transformations, and inferences. International Journal on Document Analysis and Recognition 7(1), 1–16 (2004)
85. Zhai, Y., Liu, B.: Web data extraction based on partial tree alignment. In: Proceedings of the 14th international Conference on World Wide Web, WWW 2005, Chiba, Japan, May 10-14, pp. 76–85. ACM, New York (2005)
86. Zhai, Y., Liu, B.: Structured Data Extraction from the Web Based on Partial Tree Alignment. IEEE Transaction on Knowledge and Data Engineering 18(12), 1614–1628 (2006)
87. Zhang, R.Y., Lakshmanan, L.V.S., Zamar, R.H.: Extracting relational data from HTML repositories. ACM SIGKDD Explorations Newsletter 6(2), 5–13 (2004)
88. Zheng, S., Song, R., Wen, J.-R.: Template-independent news extraction based on visual consistency. In: Proceedings of AAAI–2007, pp. 1507–1511 (2007)
89. Zheng, S., Zhou, D., Li, J., Giles, C.L.: Extracting Author Meta-Data from Web Using Visual Features. In: Data Mining Workshops, ICDM Workshops, pp. 33–40 (2007)
90. Zhu, J., Nie, Z., Wen, J., Zhang, B., Ma, W.: Simultaneous record detection and attribute labeling in web data extraction. In: Proceedings of the 12th ACM SIGKDD international Conference on Knowledge Discovery and Data Mining, KDD 2006, Philadelphia, PA, USA, August 20 - 23, pp. 494–503. ACM, New York (2006)
91. Zhu, J., Zhang, B., Nie, Z., Wen, J.R., Hon, H.W.: Webpage understanding: an integrated approach. In: Conference on Knowledge Discovery in Data, San Jose, California, USA, pp. 903–912 (2007)
92. Zhu, M., Hu, W.: Topic Detection Tracking for Threaded Discussion Communities. In: International Conferences on Web Intelligence and Intelligent Agent Technology, Sydney, Australia, pp. 77–83 (2008)
93. Zou, J., Le, D., Thoma, G.R.: Combining DOM tree and geometric layout analysis for online medical journal article segmentation. In: Proceedings of the 6th ACM/IEEE-CS Joint Conference on Digital Libraries, JCDL 2006, Chapel Hill, NC, USA, June 11-15, pp. 119–128. ACM, New York (2006)

Chapter 5
Web Structure Mining

Ricardo Baeza-Yates* and Paolo Boldi

Abstract. This chapter covers the basic properties, concepts and models of the Web graph, as well as the main link ranking and Web page clustering algorithms. We also address important algorithmic issues such as streaming computation on graphs and web graph compression.

5.1   Introduction

Although data mining and information retrieval pre-date the birth of the Web, and of Internet itself, by many decades (the basic idea can be traced back to Vannevar Bush's much celebrated paper [26]), they have been receiving a novel impetus in the last decades because of the new challenges related to the Web and to the construction of search engines: before the advent of the Web, most users simply did not have access to large document collections and thus they did not need any complex system to help them retrieve a specific document; information retrieval tools were only used in particular circumstances or within certain professional activities.

   With the arrival of the Web, every user has gained instantaneous access to an enormous, almost infinite, and ever-growing repository of information and data of all kinds and about every possible topic: sorting out and searching this humongous amount of knowledge became a basic need and started presenting new unseen problems and directions to the field of information retrieval. These new frontiers actually

Ricardo Baeza-Yates
Yahoo! Research,
Barcelona, Spain
e-mail: `rbaeza@acm.org`

Paolo Boldi
Dip. di Scienze dell'Informazione,
Univ. degli Studi di Milano, Italy
e-mail: `boldi@dsi.unimi.it`

* Corresponding author.

J.D. Velásquez and L.C. Jain (Eds.): Advanced Techniques in Web Intelligence – 1, SCI 311, pp. 113–142.
springerlink.com   © Springer-Verlag Berlin Heidelberg 2010

shaped a new research field, sometimes referred to as *Web information retrieval* [7] to distinguish it from "traditional (or classical) information retrieval", that covers a wealth of new data mining problems. We are here understanding that information retrieval offers the basic tools used by data miners to access the data; as a consequence, it is not possible to speak of data mining without considering in detail which information retrieval primitives are available.

We can outline the differences between Web (or modern) and traditional information retrieval as follows:

- the number of documents is larger by some orders of magnitude;
- documents are extremely heterogeneous: they have different content types (text, images, audio etc.), different representation formats, different authors, different authoritativeness, different degrees of reliability (may even be adversarial!) and so on;
- documents are assumed to be contained in a single unit although some pages may contain multiple topics or a document may span multiple pages;
- documents are continuously modified: new documents are added, old documents are deleted, and some documents change their content; and
- documents are endowed with a hypertextual structure that links them to one another, as well as an internal structure expressed usually in some XML based language.

Most of these features can be seen as odds for the work of the miner: for example, having to do with a huge number of documents makes many classical algorithms and data structures almost useless and require sometimes to redesign many approaches from scratch. Not so for the presence of hypertextual links, though: the fact that documents link to one another introduces a new albeit somehow implicit source of information, and enriches the data. Probably the best known among all Web-related algorithms, PageRank [66], is based solely on the linkage information. Understanding the meaning behind the links and exploiting them to get new insight on the documents is at the base of Web Structure Mining, the topic we are going to discuss in this chapter.

5.2   The Web as a Graph: Facts, Myths, and Traps

Before proceeding, let us establish some basic notations, definitions and conventions that we will be using throughout this chapter. A *(directed) graph* (or "digraph") $G = (V, E)$ is defined by a set of *nodes* or *vertices* V and a set of *arcs* or *edges* $E \subseteq V \times V$; since in most of this chapter we are going to consider directed graphs, we will usually omit the adjective "directed". We write $n = |V|$ for the number of nodes and $m = |E|$ for the number of arcs. The graph $G^T = (V, E^T)$ with $E^T = \{(y,x) \mid (x,y) \in E\}$ is called the *transpose of G*.

Given a node $x \in V$, we let $N_G^+(x) = \{y \mid (x,y) \in E\}$ (the set of *out-neighbors*) and $N_G^-(x) = \{y \mid (y,x) \in E\}$ (the set of *in-neighbors*); the values $d_G^+(x) = |N_G^+(x)|$ and $d_G^-(x) = |N_G^-(x)|$ are called the *out-degree* and *in-degree* of x, respectively. The subscript G will be omitted when clear from the context. A *path* is a sequence of nodes $\pi = (x_1, x_2, \ldots, x_n)$ such that $(x_i, x_{i+1}) \in E$ for all $i < n$; we say that π has length $|\pi| = n - 1$, *starts* at x_1 and *ends* at x_n; the set of all paths from x to y is denoted by $\Pi_G(x, y)$. The path π is *simple* if it contains no repeated nodes. It is a *cycle* if $x_1 = x_n$; a simple cycle is a cycle that contains no repeated nodes (except for the fact that $x_1 = x_n$).

We say that y is *reachable* from x in G (written $x \rightarrow_G^* y$) iff there is a path from x to y, i.e., $\Pi_G(x, y) \neq \emptyset$. The equivalence relation on nodes defined by letting two nodes x and y be equivalent iff $x \rightarrow_G^* y$ and $y \rightarrow_G^* x$ partitions the set of nodes into equivalence classes that are called the *strongly connected components (SCCs)* of G; we say that G is *strongly connected* iff it has but one SCC.

An *undirected graph* is a graph $G = (V, E)$ such that[1] $(x, y) \in E$ implies $(y, x) \in E$. In an undirected graph, out- and in-neighbors and degree are simply called *neighbors* and *degree* (and denoted by $N_G(x)$ and $d_G(x)$, respectively). Every directed graph G may be seen as undirected, augmenting it with all the missing reciprocal arcs: this is called the *undirected version of G*.

If G is undirected, then $x \rightarrow_G^* y$ implies $y \rightarrow_G^* x$, so there can be no edges between nodes lying in different components.

In most cases, we want to treat graphs *up to isomorphism*; in other words, we consider that two graphs (V_1, E_1) and (V_2, E_2) are the same iff there is a bijection $\phi : V_1 \rightarrow V_2$ such that

$$(x, y) \in E_1 \Longleftrightarrow (\phi(x), \phi(y)) \in E_2.$$

Hence, we can always assume that the node set V is $V = \{1, 2, \ldots, n\}$: unless stated otherwise, in the following we will always take this assumption for granted.

Given a graph $G = (V, E)$, its *adjacency matrix* $A = A_G$ is the square matrix of size $n = |V|$ such that

$$a_{ij} = \begin{cases} 1 & \text{if } (i, j) \in E \\ 0 & \text{otherwise.} \end{cases}$$

5.2.1   The Web Graph

The *Web graph* is the graph whose nodes are the URLs and that contains an arc (x, y) iff the page identified by the URL x contains a hypertextual link to y. This seemingly innocent definition contains a number of nontrivial implicit surmises, that we want to highlight briefly in the following; some of these points will be later dealt with in more detail.

- *Dependence on time.* The Web is an ever-changing, intangible object: at every moment in time new URLs are added, deleted, and pages are modified. From the seminal papers [29, 65], it is clear that about 5% of new content is introduced in

---

[1] Technically, this is usually called a "symmetric digraph" in the literature: most textbooks on graphs prefer to define an undirected graph as one whose arcs are unordered pairs of nodes. For the purpose of this chapter, it is more convenient to identify symmetric digraphs and undirected graphs (the difference only concerns the treatment of self-loops).

the Web every week, and about 80% of the links change; these data have been recently confirmed by the first large-scale study in this regard [22]. The Web graph is also difficult to capture: a specific graph would be obtained by taking an instantaneous snapshot of the Web at a certain moment. Albeit possible in principle, clearly such a snapshot cannot be obtained in practice.

- *Which URLs?* Even assuming that we are freezing the Web in time, it is not at all clear what the set of URLs should be. A URL is just a string denoting a resource available somehow and somewhere; there is no such thing as a URL list that contains the set of all existing URLs. As a mental process, if for a moment we define a *valid URL* as a syntactically correct HTTP URL string for which the corresponding server is providing a valid answer (a 2xx result [40]), then the set of valid URLs would certainly be infinite, and even if it was finite, collecting it would be out of question, both because there is no public list of HTTP servers, and because a server does not publish any list of the valid URLs it is providing. URLs can only be discovered through a gathering process, which is carried out by some crawler, starting from a set of well-known URLs (the *seeds*) and stopping the process when some criterion is satisfied. Even disregarding time dependence, this process is arbitrary in many respects, making the very notion of Web graph chimeric. In practice, one has to content himself with a *fixed set of URLs* that have been already crawled somehow (so that we have their content at our disposal).
- *Which links?* In the discussion above, and also in the rest of this chapter, we tried to distinguish between a URL and its content (the latter being referred to as a "page"). The result of a crawl is a set of URLs and, for each such URL, a corresponding crawled page. At the level of generality of the present discussion, a page can be anything: a piece of HTML code, an image, an audio clip, a PDF document, etc. The type of the page being determined by the content-type header in the HTTP response provided by the server. Not all types of pages may contain hyperlinks, and even those that do may provide them in a form that is not easy to be extracted automatically. Think, for example, of links that are generated by Javascript, or those embedded in Flash code [43]. These examples may seem contrived, but they are becoming more and more common with the widespread usage of AJAX-like techniques [45].
- *Redirects, normalization and DUST.* A further difficulty in the treatment of links is related to the presence of *redirects*: a server may respond to a requested URL x redirecting to another URL y (either providing a 3xx response code together with the location header [41], or using a META refresh element [74]). We may either consider x as a node of the graph with a single out-neighbor y, or we may quotient the graph with respect to the redirection relation. A similar situation happens when the same URL is presented in slightly different way, that are equivalent up to *normalization*; different notions of normalization may give rise to different graphs. A related but markedly different scenario is when different URLs produce the same content, typically because of the presence of session IDs or other immaterial arguments in the text of the URL; this is sometimes referred to as *DUST* [10] (Different URLs with Similar Text).

The considerations above suggest that there is no such thing as *the* Web graph; one should rather speak of different graphs each representing a portion of the Web.

5.2.2   Some Structural Properties of the Web

The first paper that tried to study the structure of the Web at large was [23]. The main conclusions of that research is that the Web graph has a peculiar bow-tie structure, as shown in Figure 5.1.

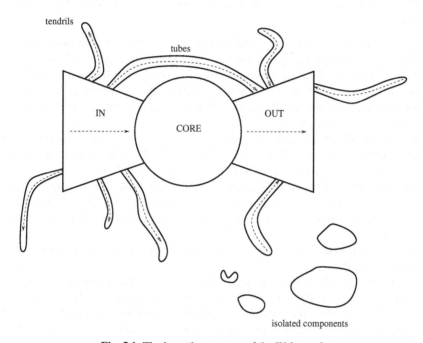

Fig. 5.1 The bow-tie structure of the Web graph

The sizes of its strongly connected components follow a power law distribution, with the largest strongly connected component (called the *Web core*) comprising about 26% of all nodes; the Web core is further reachable by another 21% of the remaining nodes (the *IN*), and about as many nodes are reachable from the core (the *OUT*). The remaining part of the Web is either totally disconnected from the rest, or forms tubes connecting IN and OUT, or tendrils attached to IN/OUT. It is a widely accepted fact that all important Web pages are in the Web core, but one should not underestimate the relevance of OUT, both because of the presence of documents (i.e., pages that do not contain links, like most of the PDF or PostScript documents) and because of the so-called Web frontier [34] (the set of documents whose out-neighbors were not crawled, due to the limits in the scope or size of the crawl performed).

As for IN, its overall status (and existence) is elusive: its presence can only be determined using a rich set of seeds (for example, starting with a single seed in the Web core would lead to an empty IN set), and its size seems to depend more on the way the seeds are chosen and the crawl is performed than on actual structural properties of the Web as a whole.

The overall structure described in [23] has been confirmed in more recent papers sometimes considering only smaller regional subgraphs [32, 78, 27, 6], although sometimes the relative sizes of IN/OUT/Core have been estimated differently probably for the reasons discussed above.

Other properties that have been observed consistently in all Web graphs are the following:

- the diameter (i.e., the length of the longest shortest path between two nodes) in the core is small with respect to the size and density of the core: [23] reported it to be 28; the length of the longest shortest path in the *whole* graph seems to be much longer (larger than 500, according to [23]); albeit smaller than in a random graph, these data contrast the folklore belief that any two pages are just "a few clicks apart" from each other;
- the average distance (i.e., the expected length of a shortest path) was estimated to be 19 in [4], whereas [23] gives a value of 16; if the undirected version of the graph is considered, the average distance becomes about 7;
- the distribution of in-degrees follows a power law with an exponent between 1.7 and 2.1; the average degree was estimated to be around 8 [23], but more recent studies suggest that the actual value may be as large as 30, and constantly growing [22];
- also the distribution of out-degrees follows a power law, with an exponent between 1.9 and 2.7;
- as for the size of the whole Web graph, the most recent estimate is of at least 11.5 billion pages (as of 2005) for the indexable part of the web (i.e., disregarding the so-called *dark Web*, the part of the Web that cannot be accessed by standard search engines because it cannot be automatically crawled: the hidden or dark Web is reported to be 500 [sic!] times larger than the indexable web [12]).
- most Web sites are not connected to the central component of the Web and are dispersed in small connected components called islands. As it is not easy to crawl all islands, not many studies give their real number. In [9] the structure of the Chilean Web is studied, where from all .cl domain names, more than 50% of the Web sites are islands. The same paper studies the dynamics of the structure showing that it is very chaotic and only a small percentage is stable.

In [32], the authors perform a more fine-grained study of the inner structure of the various parts of the Web, and conclude that although different portions of the Web (e.g., national domains [6]) feature similar properties and exhibit analogous structures, this self-similarity does not hold when individual components are considered: for example, the IN and OUT parts of a Web graph do not themselves reveal a comparable bow-tie structure.

The observed phenomena, albeit important, should be taken with a grain of salt: as we tried to outline above, there is a number of elements that influence the shape of the graph (the most important being the behavior of the crawler used to download the data), and for every single feature one should carefully judge whether it is a property of the Web graph itself or whether it is determined as an artifact of the choices operated while collecting it.

As a final note, like with Web mining in general, data availability is a central issue in the study of Web graphs: apart for commercial search engines (that are clearly reluctant in releasing large data sets), it is rare that a single research group has the means and infrastructure needed to collect significant amounts of data. Nonetheless, differently from other kinds of Web data, once collected a graph can be easily stored, communicated and made available.

5.2.3   Web Graph Models

As in most data mining scenarios, the search for significant patterns in the data requires one to compare them against "typical" patterns, whose presence is expected and thus uninteresting. It is not surprising, therefore, that one of the most developed and active areas of research is that concerning Web graph models. Formally, a *graph model* is a family of distributions of graphs, one for each graph size, that usually depends on a number of parameters and takes the form of a generating process. A graph model is supposed to produce typical graphs found in practice: in the case of Web graphs, we expect a "good" Web graph model to produce with high probability graphs that are similar to real-world Web graphs, as described above.

The usage of Web graph models is twofold: on one hand, it allows one to produce synthetic graphs that "look like" a Web graph on which algorithms can be experimented without the need of actually collecting the data (and more importantly: properties of those algorithms may be proved or disproved formally under the given model); on the other hand, the availability of such models allows one to determine how likely it is to observe a certain pattern, and this fact may be used to spot anomalies and outliers.

As we said, most graph models take the form of a generative process: a sequence of graphs $G_1, G_2, \dots$ is produced; at every step, new nodes and/or arcs are added to G_n, leading to G_{n+1}: the process of adding nodes and arcs is probabilistic. When a graph of the desired size is obtained, the process stops. Some models produce undirected graphs, while other are designed for directed graphs: in the case of Web graphs, the former one is used when one just wants to produce the undirected version of a Web graph.

The theory of graph models dates back to the late 50's; in particular, the fundamental model, called the *Erdős-Rényi graph model* $\mathscr{G}_p$ (sometimes dubbed "random graph model"), introduced independently in [46] and [37], generates undirected graphs and depends on a single parameter $p \in [0, 1]$. This generative model adds one vertex at a time, starting from the one-vertex graph G_1; after a new vertex $n + 1$ has been added, it is made adjacent to each of the existing n vertices with probability p, independently for every vertex.

This model induces a probability distribution where the probability of a given graph G only depends on the number of vertices and edges of G, and is precisely given by

$$P_{\mathscr{G}_p}(G) = p^m (1-p)^{\binom{n}{2}-m}.$$

An alternative, non-generative, model $\mathscr{G}_m$ also proposed by Erdős and Rényi, has the number of edges m as unique parameter, and produces the graph by choosing the set of m edges uniformly at random among all the possible sets of edges of cardinality m. It is not difficult to see that, by the law of large numbers, almost all graphs in $\mathscr{G}_p$ will have $m = pn(n+1)/2$ edges, which makes the two models almost interchangeable.

Although apparently simple, the Erdős-Rényi model features a number of interesting non-trivial properties, and can be fruitfully used to produce graphs that are similar to those that appear in a number of practical circumstances, but it is unfortunately unsuitable in the case of Web graph. For example, it is not hard to see that the distribution of node degrees is a binomial, and not a power law; more precisely, the probability that a node has degree k is

$$\binom{n-1}{k} p^k (1-p)^{n-1-k}.$$

This observation prompted the need for developing new, more appropriate models for Web graphs.

5.2.3.1 Preferential-Attachment Models

The first family of Web graph models is based on the idea of choosing the target of each new edge not uniformly among the existing vertices but rather proportionally to their current degree: these models are often referred to as *preferential attachment*. The obvious interpretation of these models is that pages that are popular tend to be linked to more often, thus becoming even more popular; for this reason, the preferential attachment paradigm is sometimes called *rich-get-richer*.

One example of preferential-attachment model, defined in [20], is called *LCD* (for "Linearized Chord Diagram") and depends on a single integer parameter m. At every step, a single new vertex is attached, with m new edges, each joining the new vertex to one of the existing vertices; for every edge, the endpoint is chosen with a probability proportional to its current degree. The LCD model can be used to generate directed graphs (in this case, the new arc always starts from the new node and ends into one of the existing nodes).

As proved in [20] and [19], the graphs generated according to the LCD model enjoy a number of properties that are typical of small-world graphs like the Web: for example, almost certainly (i.e., with probability converging to 1 as $n \to \infty$) the diameter of the graph is about $\log n / \log\log n$, as long as $m > 1$. Also, degrees (in the directed case: out-degrees) are distributed according to a power law; unfortunately, the induced power law distribution has a fixed exponent 3 (independently on the choice of m), which is markedly different from the empirical results on real-world Web graphs.

In [3], another preferential-attachment model is defined, based on a single parameter $\alpha \in (0,1)$; in this model, at each step either a single arc (with probability α) or a single node (with probability $1 - \alpha$) is added: when an arc is added, its starting (ending, respectively) node is chosen proportionally to the out-degrees (in-degrees, resp.)[2]. It can be proved that the in- and out-degree distributions both follow a power law with exponent $1 + 1/\alpha$, so α can be tuned to obtain any desired power law exponent in the range $(2, +\infty)$. Albeit more reasonable, this is still slightly unsatisfactory, in that one cannot differentiate the in- and out-degree distribution from each other.

A more sophisticated variant of the latter model, also proposed in [3], introduces two new parameters γ^- and γ^+, and the starting (ending, respectively) node of a new arc is chosen to be x with probability proportional to $\gamma^+ + d_G^+(x)$ ($\gamma^- + d_G^-(x)$, resp.). The resulting in- and out-degree distributions follow two power laws with exponents β^+ and β^-, where

$$\beta^\pm = 2 + \frac{1-\alpha}{\alpha}\gamma^\pm$$

This variant is more versatile and allows one to de-couple the two degree distributions, producing much more realistic graphs.

For other models based on preferential attachment, we invite the reader to look at [33].

5.2.3.2   Copying Models

Another family of Web graph models are those based on *copying*: when a new node is created, it copies its content (hence, in particular, its out-links) from another existing node. Also these models mimics a well-known behavior of page editors, that often tend to copy-and-paste to create new pages. An example of copying model is the *linear growth copying model*, introduced by Kleinberg *et al* in [31].

The linear growth copying model depends on two parameters: a positive integer d and a further threshold $p \in (0,1)$. The generative process starts from a directed graph H (whose nodes have all out-degree d), and adds one node with d outgoing arcs at every step. Whenever a new node x is added, an existing node y is chosen uniformly at random: for each of the d out-neighbors z of y, with probability p the arc (x,z) is added; at the end of this phase, the node x will have out-degree $d' \leq d$, and $d - d'$ further arcs are added from x to $d - d'$ existing nodes chosen uniformly at random.

Although the graphs produced by this model have constant out-degree d, their in-degree distribution follows a power law, with exponent

---

[2] More precisely, the node x is chosen as starting (ending, resp.) node with a probability proportional to $d_G^+(x) + 1$ ($d_G^-(x) + 1$, resp.), where G is the current graph.

$$1 + \frac{1}{1-p}.$$

More complex variants of this model allow one to obtain more versatile distributions also on the out-degrees (see, e.g., [21]).

5.3   Link Analysis

As we highlighted, the presence of hyperlinks is a unique feature of the Web, and it is thus natural to try to exploit it as a further source of information about Web pages. To understand this point better, it is worth to distinguish between two types of links: navigational and informational. *Navigational links* are there to allow the reader to navigate the pages of a Web site (e.g., "home", "up", "next" etc.); *informational links* refer the reader to other pages, possibly belonging to a different Web site, that are deemed important (interesting, useful, etc.) for the reader to consult.

Clearly, these two types of links provide different kinds of information, and they should be treated separately; in particular, the links that are more interesting from our viewpoint are the informational ones. Unfortunately, there is no automatic way to distinguish between the two types of links, so most authors choose one of the following approaches:

- some classifier is adopted that is able to approximately determine the type of a link: the classifier may use information such as the textual content of the hyperlink (the so-called "anchor text"), the actual URL that is pointed to, etc.;
- all links that point to a URL belonging to the same host (sometimes referred to as *nepotistic links*) are considered to be navigational, whereas the others are informational; or
- all links are considered as informational.

From now on, we shall assume that only informational links have been retained, whatever method is used to discard the other ones, if any. For this reason, in the following the adjective "informational" will be dropped, and we shall simply speak of links.

A link is often compared with (and thought to be roughly equivalent to) a citation in a book or a paper; in this sense, the whole study of the Web graph is similar to what people have been doing for a long time in bibliometry [67] when trying to understand the co-citation patterns of the literature in a certain area or about a certain topic. Typically, a citation is assumed to be a way to confer importance: if x is citing y, then x's author thinks that y is important; this is clearly an oversimplification (in some cases, citations do not have any positive meaning, and sometimes they even carry a negative sentiment), but it turns out to be extremely useful in trying to understand the relevance of hyperlinked information, giving raise to a whole family of ranking methods collectively known as *link-analysis ranking (LAR)* techniques. Another way to use co-citation is to assume that if x cites y, both papers are on the same topic. Hence, if z also cites y, we could assume that x and z are related (see for example [49]).

LAR techniques all aim at using the linkage structure to deduce the importance of Web pages. The main advantage of such methods over other more classical ones stands in the fact that the information is attributed exogenously, and not endogenously, leading to a system that is supposedly harder to manipulate maliciously.

In a very broad sense, a LAR method assigns an importance score to a set of Web documents using their linkage information. In this general scenario, the linkage information (i.e., the graph) is possibly used *along with* other data (e.g., the actual content of Web pages, or other external source of knowledge): when only the graph is used, we speak of *pure LAR*, as opposed to *hybrid LAR*. Sometimes (like in the HITS algorithm that we detail later [57]), the technique uses external information only to select a subgraph, after which the score is computed using solely the subgraph itself (the nodes outside the selected subgraph are not relevant, and they can be assigned the lowest possible score).

In this chapter we only discuss pure LAR techniques, or techniques that can be reduced to pure LAR as explained above. Such a technique $\mathscr{A}$, given a graph G with n nodes, produces a *score*[3] $\sigma_G^{\mathscr{A}} : V_G \to [0,1]$. We shall omit $\mathscr{A}$ and/or G whenever it is clear from the context.

The simplest kind of LAR method consists in counting the number of incoming links, an idea that was suggested in many early LAR papers [63, 52, 62]. That is, letting

$$\sigma_G(-) = d_G^-(-).$$

This is the *in-degree ranking*, and corresponds to the natural idea of citation counting: the importance of a page is the number of pages that point to it. To normalize scores, we actually have to modify slightly our definition letting

$$\sigma_G(x) = \begin{cases} d_G^-(x)/\vartheta & \text{if } d_G^-(x) \le \vartheta \\ 1 & \text{otherwise} \end{cases}$$

where ϑ is a threshold on the number of incoming links we want to consider. Albeit simple and although often accused to be too easy to spam, in-degree works surprisingly well in many cases [73].

5.3.1 PageRank

A more sophisticated LAR technique is PageRank and will be described in this section. PageRank [66] is one of the most well-known measures of importance of a web page: inspired by previous works on the mutual citations for determining the relevance of scientific papers, it is based on the intuition that a web page is more important if it is linked to by many important pages. PageRank is one of the factors used by search engines to determine the order of answers to a query, a problem of uttermost importance that is often referred to as *web ranking*, hence the name "PageRank".

---

[3] Without loss of generality, we assume the score to be normalized to the unit interval.

One suggestive metaphor to describe the idea behind PageRank is the following: consider an iterative process where every web page has a certain amount of money that will at the end be proportional to its importance. Initially, all pages are given the same amount of money. Then, at each step, every page gives away all of its money to the pages it points to, distributing it equally among them: this corresponds to the interpretation of links as a way to confer importance. This idea has a limit, however, because there might exist groups of pages that "suck away" money from the system without ever returning it back. Since we want to disallow the creation of such oligopolies, we force every page to give a fixed fraction $1 - \alpha$ of its money to the State (like tax). The money collected this way is then redistributed among all the pages either equally or according to some criterion, represented as a vector v whose i-th component is the fraction of money that will be given back to page i. Still, pages that "suck away" money may need to be handled differently and also we should remove self-links (that is, money going back to the same page).

In the PageRank jargon, the vector v is called the *preference vector* (or teleportation vector), whereas α is called the *damping factor* (and $1 - \alpha$ is sometimes called the teleportation probability).

To define PageRank formally, let[4]

$$P = \text{Diag}(G\mathbf{1})^{-1}G.$$

Some care must be exercised here, because of the presence of *dangling nodes*, i.e., nodes with no out-neighbors: indeed, the above definition of P is well-given only if no dangling node is present. We postpone the discussion about the presence of dangling nodes; for the time being, we assume that G contains no dangling nodes.

Let now

$$M = \alpha P + (1 - \alpha)\mathbf{1}v^T.$$

The matrix P is row-stochastic, and so is M, which also turns out to be unichain [15], so it has a unique invariant distribution, i.e., there is a unique stochastic vector r such that

$$r^T M = r^T.$$

Such vector is precisely PageRank, as informally described above. Notice that this can be interpreted as an eigenvector equation. That is, the solution is the eigenvector associated to the eigenvalue 1 of the matrix M (this eigenvector always exist for a stochastic matrix). Substituting the definition of M in the above formula, we have

$$\alpha r^T P + (1 - \alpha)r^T \mathbf{1}v^T = r^T;$$

since $r^T \mathbf{1} = 1$, we have

$$\alpha r^T P + (1 - \alpha)v^T = r^T.$$

---

[4] All vectors are column vectors; $\mathbf{1}$ is a vector of 1's, and $\text{Diag}(v)$ is the diagonal matrix with trace v.

We can solve r^T obtaining

$$r^T = (1 - \alpha)v^T(I - \alpha P)^{-1}.$$

From the previous formulation we have that for any given node i

$$r_i = \alpha \sum_{j \in N_G^-(i)} \frac{r_j}{d_G^+(j)} + (1 - \alpha)v_i.$$

This formula has a natural interpretation: the importance of page i depends partly on v_i (the preference of the page i, according to the user) and partly on the importance of the pages that point to i ($N_G^-(i)$); each of them distributes its own importance (r_j) equally among the pages it points to. Notice that this formula is recursive in its nature, and the solution may be found exploiting methods from linear algebra.

As observed many times (see, e.g., [5]), an equivalent definition is the following

$$r_i = \frac{1}{1 - \alpha} \sum_j \sum_{\pi \in \Pi_G(j,i)} \alpha^{|\pi|} v_j b(\pi)$$

where, $b(\pi)$ (the *branching contribution* of π) is defined as follows

$$b(\pi) = \frac{1}{d_G^+(x_1) \dots d_G^+(x_{n-1})} \quad \text{if } \pi = (x_1, \dots, x_n).$$

In practice, every path incoming in node i contributes to its importance. The value of the contribution is proportional to the preference of the starting node of the path, inversely proportional to the product of the degrees of the nodes along the path, and exponentially decays with the length of the path.

Some observations are in order, here:

- *Computation and approximation.* The closed formulae presented for PageRank are of little help in practice. The methods that are used for actually computing the PageRank scores are typically iterative techniques employed in linear algebra for solving linear systems or for eigenvalue problems, PageRank being the principal norm-1 eigenvector of M. The method suggested in [66] was the classical Power Iteration method, but other alternative techniques are now available [55, 54, 59] that are more efficient, more easily parallelizable and/or apt to distributed computation, or more stable from the numerical viewpoint. Notice that all iterative techniques, however, do not actually provide the PageRank *values*, but rather an approximation with an error depending on the number of iterations performed. On the other hand, what really matters when ranking pages is the order of the ranking and not the exact values. So if the error is small enough to keep the order unchanged, we may already stop.
- *Convergence speed.* The convergence speed of all iterative methods described above depends crucially on α; while at $\alpha = 0$ PageRank coincides with the

preference vector v, when α gets close to 1, convergence is slower and slower, and the problem itself becomes numerically unstable [58].

- *Treatment of dangling nodes.* The presence of dangling nodes cannot be ignored: a large fraction of nodes are dangling, either because they are documents (most PDF, PostScript files and so on do not contain hyperlinks) or because they fall in the frontier of the crawl [34]. As observed in [16], the way dangling nodes are treated changes crucially the results obtained. The most reasonable solution consists in putting a copy of v in M at every row corresponding to a dangling node: this is called *strongly preferential* PageRank in [16].

- *Topic-sensitive PageRank.* In the definition of PageRank, the rôle of the preference vector v is critical, and it intuitively corresponds to the profile of the user that is performing the search. One may want to think of computing one different PageRank vector for every user, depending on her preferences, but this is of course infeasible. A more reasonable solution, proposed in [48], is that of computing a restricted (small) number of preferential PageRank vectors, one for each "topic", and in combining them linearly depending on weights that are determined, for example, by the user's profile or history. Many recent works follow this idea, proposing personalized variants of PageRank (see e.g. [44]).

- *Generalizing PageRank.* The path formula given above can be seen as a form of weak transitivity of importance, with α controlling the decay of importance as it flows along paths; such decay is exponential, but it can be sensible to consider other forms of decay that may fit better one's intuition about the network at hand. Such a form of generalized PageRank was discussed in [8].

5.3.2   HITS

In [57], an alternative LAR method was proposed known as HITS; the idea behind HITS is that every page has two scores, rather than one: the first score a_i measures how authoritative page i is (i.e., how valuable or reliable is the information it contains), whereas the second score h_i measures how much page i is a hub (i.e., a container of authoritative links). Clearly, an authoritative page is one that is pointed by many hubs, and a good hub is one that points to many authoritative pages.

This amounts to requiring that

$$a_i = \sum_{j \in N_G^-(i)} h_j$$
$$h_j = \sum_{i \in N_G^+(j)} a_i.$$

This request can be turned into an iterative algorithm, keeping in mind that convergence is not directly guaranteed and must rather be enforced at every iteration.

More precisely, given a pair of normalized score vectors (a, h), a new pair of vectors (a', h') is computed by letting

$$a'_i = \sum_{j \in N_G^-(i)} h_j$$

$$h'_j = \sum_{i \in N_G^+(j)} a'_i;$$

finally, both vectors are normalized, obtaining a new pair of normalized score vectors for the next iteration. As proved in [57], this computation converges to a pair of score vectors (a^*, h^*) that are the principal eigenvectors of $G^T G$ and GG^T, respectively, provided that the two largest eigenvalues of G have distinct modules.

The HITS algorithm itself is not usually applied directly to the whole Web graph, but rather to a subgraph depending on the query. According to the original description of the algorithm, given a query q, the set R_q of the topmost t pages matching q are considered, according to some content-dependent ranking. After that, for each node $i \in R_q$ we consider the set of all nodes pointed by i (i.e., $N_G^+(i)$) and an arbitrary set of at most d nodes that point to i (i.e., any subset of $N_G^-(i)$ of cardinality d at most). The resulting set of pages S_q is then used to build the graph on which the HITS algorithm is run.

The advantages of HITS over PageRank have often been discussed, and there is no clear reason to choose the first over the second, especially considering that HITS cannot be computed offline, and thus it is difficult to apply it in the setting of a general search engine. The interested reader may however find more information on this in [64].

5.3.3   Spam-Related LAR Algorithms

PageRank and HITS are two examples of LAR techniques aiming at establishing authoritativeness of Web pages (another similar method that is not discussed here is SALSA [60]). Links, however, may find another use that is somehow orthogonal to authoritativeness, related to Web spam.

In the context of the Web, the term *spam* refers to any activity that aims at obtaining an undeservedly large visibility on search engines. Authors of Web pages employ spam-like techniques to make their pages/sites appear more authoritative than they are, in absolute or about certain topics. Spam is not always a malicious activity, and the less derogatory expression *search-engine optimization* or SEO is preferred by many. Nonetheless, the presence of spam certainly skews the perceived reliability of Web pages and naturally requires search engines to make every possible effort to detect or demote spam pages.

Spam techniques may affect the content or the link structure of the pages involved, and here we shall particularly be interested in the second case: link spam is finalized to hoax LAR algorithms by introducing fake links or otherwise modifying the link structure of a Web site with malicious intents. Sometimes, the techniques adopted are so sophisticated that they truly require to build an artificial structure with hundreds (or thousands) of pages and links, called a *link farm*.

The presence of spam nodes, and the duality between ham[5] and spam pages, poses a natural question: can a ham node contain a hyperlink to a spam node? Clearly not, at least provided that the author of the ham node understood that the spam node was actually spam. So "spammicity" is a property that propagates in a direction contrary to that of links. This idea is at the basis a link-spam detection algorithm known as *BadRank*, that among the SEO community is believed [77] to be used by some search engines (most notably, Google). The algorithm starts from a set of pages marked as "bad" and uses a technique similar to PageRank (but with arc direction reversed) to diffuse badness.

A conceptually similar, but dual, algorithm is *TrustRank* [47]: TrustRank starts from a set of trusted nodes, and uses a PageRank-like diffusion algorithm to spread the trustiness value all over the graph. Nodes with low trust value are then penalized as suspect of being spam-like.

A related idea [77] is to try to individuate link farms consisting of TKC (Tightly Knit Communities), following the same route as in [14] but obtaining an algorithm that can be applied fruitfully also to large (Web-size) graphs. They deem a page "bad" iff there are too many domains that happen to be both out- and in-neighbor of the page: the set of bad pages is called "bad seed". Then, the bad seed is expanded recursively according to the idea that a page that points to too many bad pages is itself bad, as in BadRank. Finally, PageRank is computed penalizing or deleting links from bad sites.

5.4   Structural Clustering and Communities

One important mining task in the Web is finding clusters of Web pages. A first approach would be a pure text based (or topical) clustering. A second approach, would be to use just the link structure among Web pages. This approach has the advantage that exploits the local structure and is more efficient than a pure text based clustering. A third approach would be to combine both, considering content as well as structure. In this case the content based part can be run after the structural clustering is done. Here we cover the second approach, which is the one in the scope of this chapter.

For this we consider a different setting which is similar to links among Web pages: a social network. Links within a social network represent some type of relations. The only difference is that links in the Web are directed while in social networks they are usually undirected. Nevertheless, direction does not add much information and can always be considered after clusters using undirected edges are obtained. In the case of social networks, it is natural to try to mine them to devise substructures, typical patterns, etc. The case of clusters is one of the most common applications, as each cluster represents a community of people. Hence, this problem is called *detection of communities*. A community can be loosely defined as a dense subgraph that is sparsely connected to the rest of the network.

---

[5] The word "ham" means "non-spam".

Algorithms to find clusters in graphs (networks) can be classified in *partitioning* and *hierarchical* techniques. Partitioning clustering techniques include k-means, neural network (e.g. self-organizing maps), multidimensional scaling (e.g. singular value decomposition and principal component analysis) and flow based algorithms. Hierarchical algorithms include divisive (e.g. based in betweeness centrality) and agglomerative (e.g. linkage clustering) clustering methods. Hierarchical algorithms do not allow communities to overlap, while other techniques based in local computations such as the k-clique percolation algorithm do. A classical algorithm that can be used to improve a given clustering is the Kernighan and Lin algorithm [56] that swaps groups of nodes trying to improve a quality function.

The main issue in all the mentioned algorithms is how to define the just mentioned quality function. One of the most popular functions is *modularity*, which measures how well different communities are divided. In the most simple case, modularity is defined as

$$Q = \sum_i \left(e_{ii} - \left(\sum_j e_{ij} \right)^2 \right)$$

where e_{ij} denotes the fraction of edges that start in community i and ends in community j. Note that $Q = 0$ if all nodes are in the same community. For weighted networks, we count weights instead of edges. Modularity is related to *assortativity*, that measures how nodes are connected to similar nodes. Another measure used is the *clustering coefficient* which is defined as

$$C = \frac{3 \text{ number of triangles}}{\text{number of triples of connected vertices}} \tag{5.1}$$

The clustering coefficient also can be defined in a local manner and then compute the global coefficient by averaging all of the local ones [76]. For more details on this topic we refer the reader to the excellent survey by Porter *et al* [68], which has more than 130 references, although misses some seminal algorithmical papers such as [42].

5.5   Algorithmic Issues

Needless to say, the web graph is a huge object to deal with: it currently contains some 3 billion nodes, and more than 50 billion arcs (and these estimates are just lower bounds, as they are obtained from search engines, which index just a part of the Web). This purely quantitative observation has some nontrivial consequences not only on the kind of algorithms that can be effectively run on the graph, but also about the way the graph is actually stored: the standard adjacency matrix of a 3 billion node graph has $9 \cdot 10^{18}$ entries; even if we use just one bit to store each one of them, we would end up using more than one million terabytes of memory just to store the graph!

Of course, the Web graph itself is very sparse, so it would be wiser to represent it through its adjacency lists. With an average out-degree of 30, the concatenation of all adjacency lists would require about 330 GBytes of data. This is not much for nowadays external storage, but it still largely exceeds the size of core memories. Whether this is a serious limit or not depends on the kind of algorithms that we want to implement. For example, since PageRank requires just a linear scan of the graph for every iteration, having the graph available offline is not a major issue (but one should not underestimate the memory needed to store the rank vector: using single-precision floats we would still need about 11 GBytes for the vector alone).

These observations make it necessary to develop a practical and substantially new approach to deal with massive (Web) graphs. Two solutions have been studied in the literature: one possibility is to assume that the graph is never actually loaded into memory, but it is rather read (possibly more than once) in a streaming fashion using a small amount of memory. The second is to rely on some graph compression techniques that allow one to load the graph in memory and to access it without the need of decompressing but a small portion of it. We will briefly discuss these two combinable solutions in the forthcoming sections.

5.5.1   *Streaming and Semistreaming Computation Models*

Streaming is an important computation model on massive data sets [50]: roughly speaking, in a streaming model the data are read in a sequential (although arbitrary) order and with not enough memory to store them. More precisely, in the *graph streaming model* the computation is performed in $O(1)$ passes, and each pass consists in a sequential read of the arcs (in unknown order) using an amount of memory that is poly-logarithmic in the number of arcs m; the time needed to process each arc is usually assumed to be $O(1)$. Although a number of graph problems have been shown to admit an exact or approximate streaming solution, the model itself is too restrictive for most applications: for example, even determining whether there is a path of length 2 between two vertices (the so-called length-2 path problem) cannot be solved in the streaming model. This is due to the fact that deciding whether two sets are disjoint requires linear space [53], and determining if there is a length-2 path between two nodes x and y is equivalent to establishing whether the intersection of the out-neighborhood of x and the in-neighborhood of y is disjoint.

To overcome the restrictions that the streaming model implies, many relaxed versions have been considered:

* models that also have an output stream that they can write [30];
* models that introduce a further sorting primitive [2];
* models that relax the restriction on the amount of memory available.

In particular, in the last variant [39, 30], known as *graph semi-streaming model*, the memory available is $O(n \log^k n)$: in other words, there is not enough memory to store the whole graph, but still there is enough memory to store its nodes. There is a variety of interesting problems that can be solved in the semistreaming model: for example, there exists a $\log n / \log \log n$ approximation semistreaming algorithm for computing the diameter [39], and there are efficient algorithms to estimate the number of triangles adjacent to each vertex [11].

We want to briefly discuss the latter example, both to give the reader an idea of the kind of difficulties arising when dealing with large graphs using little memory and to present the useful technique of min-wise independent permutations introduced in [24]. To reduce the amount of technicality, consider an undirected graph $G = (V,E)$ and suppose that one wants to estimate the number of triangles that are incident on each vertex x, that is

$$T(x) = \frac{1}{2}|\{vw \in E : xv \in E, xw \in E\}|.$$

This problem is markedly different from that of studying the *overall* number of triangles in the graph (see equation 5.1 and [25]) and may find an immediate application to the computation of the *local clustering coefficient*, that is, the ratio between $T(x)$ and the number of possible triangles to which x could participate given its degree: this measure of local density turns out to be an important feature in applications such as spam detection or content quality. For example, in Figure 5.2 we present the empirical distribution of triangles in the undirected version of a web graph whose pages have been manually classified into spam and non-spam: the distributions in the two cases are markedly different.

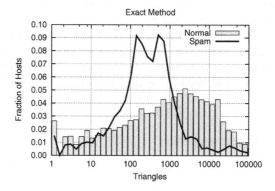

Fig. 5.2 Separation of non-spam and spam hosts in the histogram of triangles

Clearly, the number of triangles involving x can be computed by looking at how many neighbors y of x have a neighbor in common with x itself; more precisely:

$$T(x) = \frac{1}{2} \sum_{y \in N(x)} |N(y) \cap N(x)|.$$

The *Jaccard coefficient* [51] is a measure of the overlap of two sets A and B, defined as

$$J(A,B) = \frac{|A \cap B|}{|A \cup B|}.$$

Simple algebraic manipulation shows that

$$|A \cap B| = \frac{|A| + |B|}{1 + 1/J(A,B)}.$$

So, to give an estimate to $T(x)$ one just needs to have a reasonable way to estimate the Jaccard coefficient of two neighborhoods, given that

$$T(x) = \frac{1}{2} \sum_{y \in N(x)} \frac{d(x) + d(y)}{1 + 1/J(N(x),N(y))}.$$

The general idea behind the technique of min-wise permutations gives a nice way to estimate the Jaccard coefficient, that we shall roughly describe in the following. Suppose you have two sets $A,B \subseteq \Omega$ and you want to estimate $J(A,B)$; draw a random injective labelling function $\pi : \Omega \to \{1,\dots,|\Omega|\}$ and apply it to both A and B: what is the probability that the two minimum labels for A and B coincide? Clearly, if A and B are completely disjoint, it will never happen that the two minimum labels coincide (simply because the *sets of labels* assigned to the elements of A and B will themselves be disjoint). On the other hand, if A and B coincide, the minimum label will be the same with probability 1. In general, the probability that the two minimum labels will be the same is exactly $J(A,B)$. This observation provides an unbiased estimator for $J(A,B)$ that is obtained by taking N independent labelling functions and looking at how often the minimum labels coincide; general concentration results may be used to see how large we should take N to have a given probability of a certain error in the estimate. If N_m is the number of times the minimum labels turned out to coincide out of N independent experiments, we estimate $J(A,B)$ as

$$\hat{J}(A,B) = N_m/N.$$

Observe that, by simple manipulation,

$$\frac{1}{1 + \frac{1}{\hat{J}(A,B)}} = \frac{N_m}{N_m + N}.$$

Hence $T(x)$ can be estimated as

$$\hat{T}(x) = \frac{1}{2} \sum_{y \in N(x)} (d(x) + d(y)) \frac{N_m(x,y)}{N}$$

where $N_m(x,y)$ is the number of times the minimum labels of $N(x)$ and $N(y)$ coincided, out of N experiments.

It is impractical to assume that we can truly draw a labelling function at random: we shall content ourselves to have a hash function π that may not be injective;

as long as the number of conflicts is small, we shall still be able to obtain a good concentration.

Note that, in practice, we need N passes over the graph: at the i-th pass, we draw a hash function h_i, we compute for each node x the value $m_i(x) = \min_{v \in N(x)} h_i(v)$ and, for each x and y that are neighbors, we have to decide whether $m_i(x) = m_i(y)$ or not. So each pass actually requires to scan the graph twice (one to compute the m_i's and one to increment the counters $N_m(x,y)$), and needs n label variables (to store m_i) and m counters. A precise evaluation of the actual bit storage size required depends on the size of the labels and on the number of passes N, which should be chosen on the basis of the desired precision [11]. A reasonable choice is to use $\log n$ bits for each label, thus giving a memory requirement of $O(n \log n)$. The counters, that would require $O(m \log N)$ bits, are not stored in the main memory because they are only accessed sequentially[6].

5.5.2   Web Graph Compression

A completely different alternative is to use some technique that allows one to store a Web graph in memory in a limited space, exploiting the inner redundancies of the web. The overall idea behind this approach (that we loosely refer to as *Web graph compression*) is that the very nature of the Web is such that its graph may be compressed efficiently and in such a way that one can decompress only the portions that are accessed and only when they are actually accessed. Although many different compression techniques have been proposed, most of them heavily rely on the notion of universal compression code, so before proceeding we want to give the reader a brief overview of what a universal code looks like.

5.5.2.1   Brief Interlude on Universal Codes

An *instantaneous code* for the set N^+ of positive integers is a map $c : N^+ \rightarrow \{0,1\}^*$ that assigns a codeword to each number in prefix-free way (i.e., so that no codeword is a prefix of another: this is equivalent to requiring that every sequence of codewords be uniquely decodable). Of course, given any probability distribution $P(-)$ on N^+ we can determine the expected code length of c, that is

$$E_P(c) = \sum_{x \in N^+} P(x)|c(x)|.$$

The *optimal code length* E_P^* is the one that is obtained using the "best" code for that distribution; asymptotically, the optimal code length is equal to the entropy H_P of P

$$H_P = - \sum_{x \in N^+} P(x) \log P(x).$$

---

[6] A more sophisticated, albeit less precise, solution that does not require any external memory is also presented in [11], but it is not discussed here.

It is customary in this context to assume that $P(-)$ be monotonically non-increasing (i.e., small values appear "more likely" than large values). Under this restriction, a *universal code* is one that guarantees an expected code length that for every monotonic distribution is only a constant factor worse than the optimal code length E_P^*.

The advantage of using a universal code is that a universal code has reasonable compression performance guarantees *for whatever distribution*, provided that it is monotonic; so, differently from other more sophisticated coding techniques, universal codes do not require a thorough analysis of the actual input data distribution. On the other hand, if we have some information on the *actual* distribution we may choose the universal code in such a way that its expected code length for that distribution is optimal. This is the so-called *intended distribution* for the code c, and is the only distribution such that $P(x) \propto 2^{-|c(x)|}$: for that specific distribution the universal code is optimal, because the expected code length coincides with the entropy.

A further advantage of adopting universal codes instead of other (better-performing) coding strategies is that universal codes have no overhead (e.g., they do not need to store any dictionary or parameter set) and are usually extremely fast to encode/decode.

As an example, consider the following universal code, known as *Elias' γ-code* [36, 61] (or simply γ-code). The γ-code $c_\gamma(x)$ of a positive integer x is obtained by writing the binary representation $b(x)$ of x, preceded by $|b(x)| - 1$ zeroes. For example, $c_\gamma(5) = 00101$ (because the binary representation of 5 is 101, whose length is 3). Since[7] $|c_\gamma(x)| = 1 + 2\lfloor \log x \rfloor = O(2\log x)$ the expected code length of c_γ with respect to a given distribution $P(-)$ is twice its entropy, so c_γ is indeed a universal code. Its intended distribution is

$$P_\gamma(x) \propto x^{-2}$$

that is, a Zipf distribution (a type of power law distribution) with exponent 2. In other words, c_γ is optimal for encoding integers distributed as a Zipf with exponent 2, and when used for other distributions it never uses more than twice the number of bits needed by an optimal code.

A very simple, yet often extremely effective, example of universal code is the so called *nibble coding*, that is a special case of the *k-bit variable length coding* when $k = 4$. The binary representation $b(x)$ of the number x to be coded is first left-padded with zeroes to obtain a multiple of $k - 1$; then the groups of $k - 1$ bits are written, each followed by a single 0 (the "continuation bit"), and at the end a 1 is written to indicate that the value is finished. The length of the code associated with x is

$$k \left\lceil \frac{\log x}{k-1} \right\rceil = O\left(\frac{k}{k-1} \log x \right).$$

For example, the nibble coding is asymptotically $1/3$ worse than the entropy, but it is better suited for distributions that are less skewed than a Zipf distribution (because, for example, the first 2^{k-1} values all require k bits).

---

[7] Unless otherwise stated, all logarithms are in base 2.

In Table 5.1 you can see the first codewords of the γ and the nibble encoding.

Table 5.1 Comparison between the γ and the nibble universal codes

Integer	γ	nibble
1	1	1000
2	010	1001
3	011	1010
4	00100	1011
5	00101	1100
6	00110	1101
7	00111	1110
8	0001000	1111
9	0001001	00011000
10	0001010	00011001
11	0001011	00011010
12	0001100	00011011
13	0001101	00011100
14	0001110	00011101
15	0001111	00011110
16	000010000	10000111

The choice of whether it makes sense to use a universal code and of what universal code is more efficient for a given situation depends on a number of factors and, crucially, on the distribution of the data to be encoded. The latter is typically obtained by inspecting sample data with the hope of deriving a general distribution, possibly depending on as few parameters as possible. Differently from other more classical areas of compression, in the Web realm most frequently data appear to have a Zipf distribution with small exponent (typically, between 1 and 2); for such distributions, the γ-code is often the code of choice (because of its simplicity and because it is targeted towards a Zipf with exponent 2), but recently [18] new simple universal codes have been developed specifically for the situations where the exponent is smaller than 2. See also [71] for more details.

5.5.2.2 Properties of the Web Graph

One of the first attempts at using modern compression techniques for the Web graph was the Connectivity Server [13], whose authors first observed the following phenomenon: if the nodes are numbered according to the lexicographic order of URLs, adjacent nodes tend to have small differences (because most arcs are navigational, and so they are between URLs belonging to the same host, and often even in the same level of the site hierarchy). This property is called *locality* and may be exploited by storing the successor list of each node x not explicitly, but rather as a sequence of gaps from x; more cleverly, if the successors of x are $y_1 < \cdots < y_k$ we can store them as $y_1 - x, y_2 - y_1, y_3 - y_2, \ldots, y_k - y_{k-1}$ (note, by the way, that all these values are natural numbers except for the first one, that may be anyway encoded as a natural number plus a bit sign).

Such gaps are expected to be small, so they can be efficiently encoded by using some universal binary code suitable for their typical distribution. The authors of [13] used a nibble code, and an analysis of empirical data shows that the distribution of gaps is a Zipf with small exponent (around 1.2, see Figure 5.3, which is also a typical example of a power law distribution) makes the choice of nibble codes reasonable, although suitable ζ codes would turn out to have a better performance.

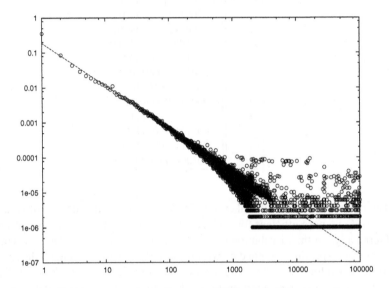

Fig. 5.3 Distribution of gaps in a typical Web graph, in loglog scale

The authors of the LINK database [70] first proposed to improve the idea of gap coding by introducing reference compression. The idea behind reference compression is that of *similarity*: URLs that are close to each other often share a large number of out-neighbors. This is because the authors of web pages tend to copy-and-paste existing pages of their own and the situation is made more and more common because of the usage of templates and of content management systems. Similarity is exploited as follows: some successor lists are encoded plainly as explained above, but some other are instead encoded differentially. When the successor list of node x is encoded differentially, another node y (called the *reference node*) is chosen that possesses many out-neighbors in common with x; a bit-mask is used to determine which successors of y are also successors of x, and the extra successors (the elements of $N_G^+(x) \setminus N_G^+(y)$) are written plainly and explicitly as explained above.

The choice of the reference node y is critical: although in principle one can choose y freely among all the nodes in the graph, for reasons of efficiency y is chosen to be one of the nodes that immediately precede x in lexicographic order. The number of predecessors of x that are considered in this choice is called the *reference window size* and it is one of the parameters that can be chosen to tune the compression performance of the system.

A further step to improve compression is related to a property called *consecutiveness*: it is often the case that the out-neighbors of a node are not only close to one another in lexicographic order, but even *consecutive*. When this happens, it makes sense to write them explicitly not one by one, but rather as an integer interval. This approach is called *intervalization* and was first used in [17]: the latter system is presently the one that achieves the best compression ratio (typically, between 2 and 3 bits/link), and a reasonable compromise between compression and speed—for this reason, it is a *de facto* standard Web graph compression format.

Some observations should be added at this point, to make the reader aware of some issues that arise in practice when compressing real-world graphs.

- When a graph compressed as described above is accessed, some decompression needs to take place. In particular, differential compression entails that accessing the successors of a node requires first to decompress fully the successors of its reference node. If the graph is scanned sequentially, the easiest way to achieve differential decompression is to keep in memory a reference window with the last successor lists (as many lists as the reference windows size). If the graph is accessed at random, though, a different solution must be adopted: decompression, in this case, happens on demand and accessing a successor list entails recursively the access of many other successor lists. To avoid unbounded recursion (which would make random access problematic) one can limit (at compression time) the length of decompression chains; of course, this choice usually reduces the compression performance, but guarantees more reasonable access time and space.

- Another subtle point to be carefully considered when a graph is accessed at random is that besides the graph itself one needs to keep track of the *offsets*, that is, the pointers to where each successor list begins. Since on average a successor list occupies about 12 bytes, offsets may turn out to require as much space as the graph. For this reason, modern implementations of compressed graph formats use some sophisticated compression techniques for the array of offsets (e.g., an Elias-Fano succinct list [35, 38]).

- Although we have only been considering abstract graphs (i.e., graphs with no labels on their nodes and/or arcs), in many practical application one needs to store additional information about the graph elements. Every compression framework should take this problem in some consideration, and should allow one to gracefully mix structure compression with label compression. Albeit important, this research area is in its infancy and results are still very immature.

- The recent trend of trying to apply the techniques and methods developed for the Web graph to other, more general kinds of networks (more importantly, the so-called "social networks" [75]) is triggering new interest in graph compression. The main limit of the compression techniques described above is that they all assume that nodes are lexicographically sorted by URL; this order (that is external to the graph) is of course not available on general graphs, and it is crucial to obtain good compression. Adapting the above frameworks to general social networks is a fast growing research area (see, e.g., [28]).

For completeness, we cannot conclude this section without mentioning that other more sophisticated approaches to Web graph compression exist that try to exploit different properties of the graph (see, for example, [1, 72, 69]), but so far they did not obtain comparable compression performances.

5.6 Summary

This chapter covers the main concepts behind mining the link structure of the Web. This includes the structural properties of the Web and the most relevant models to generate Web-like graphs, followed by the main link-based ranking algorithms such as PageRank and HITS, as well as link spam. We address also many algorithmic issues like streaming computation on graphs and web graph compression. We also cover one related problem: community detection, which is equivalent to clustering Web pages by topics.

References

1. Adler, M., Mitzenmacher, M.: Towards compressing web graphs. In: DCC 2001: Proceedings of the Data Compression Conference, p. 203. IEEE Computer Society, Washington (2001)
2. Aggarwal, G., Datar, M., Rajagopalan, S., Ruhl, M.: On the streaming model augmented with a sorting primitive. In: FOCS, pp. 540–549 (2004)
3. Aiello, W., Chung, F., Lu, L.: Random evolution in massive graphs. In: Handbook of massive data sets, pp. 97–122. Kluwer Academic Publishers, Norwell (2002)
4. Albert, R., Jeong, H., Barabasi, A.L.: The diameter of the world wide web. Nature 401, 130–131 (1999)
5. Baeza-Yates, R., Boldi, P., Castillo, C.: Generalizing pagerank: damping functions for link-based ranking algorithms. In: SIGIR 2006: Proceedings of the 29th annual International ACM SIGIR Conference on Research and development in information retrieval, pp. 308–315. ACM, New York (2006),
   http://doi.acm.org/10.1145/1148170.1148225
6. Baeza-Yates, R., Castillo, C., Efthimiadis, E.N.: Characterization of national web domains. ACM Trans. Internet Technol. 7(2), 9 (2007)
7. Baeza-Yates, R., Ribeiro-Neto, B.: Modern Information Retrieval. Addison-Wesley, Reading (1999) (Second edition will appear in 2010)
8. Baeza-Yates, R.A., Boldi, P., Castillo, C.: Generic damping functions for propagating importance in link-based ranking. Internet Mathematics 3(4), 445–478 (2007)
9. Baeza-Yates, R.A., Poblete, B.: Dynamics of the chilean web structure. Computer Networks 50(10), 1464–1473 (2006)
10. Bar-Yossef, Z., Keidar, I., Schonfeld, U.: Do not crawl in the DUST: Different URLs with similar text. ACM Trans. Web 3(1), 1–31 (2009),
    http://doi.acm.org/10.1145/1462148.1462151

11. Becchetti, L., Boldi, P., Castillo, C., Gionis, A.: Efficient semi-streaming algorithms for local triangle counting in massive graphs. In: KDD 2008: Proceeding of the 14th ACM SIGKDD International Conference on Knowledge discovery and data mining, pp. 16–24. ACM, New York (2008), http://doi.acm.org/10.1145/1401890.1401898

12. Bergman, M.K.: The deep web: Surfacing hidden value. Journal of Electronic Publishing 7(1) (2001)

13. Bharat, K., Broder, A., Henzinger, M., Kumar, P., Venkatasubramanian, S.: The Connectivity Server: Fast access to linkage information on the web. In: Proceedings of the Seventh International World–Wide Web Conference, Brisbane, Australia, pp. 469–477 (1998)

14. Bharat, K., Henzinger, M.R.: Improved algorithms for topic distillation in a hyperlinked environment. In: SIGIR 1998: Proceedings of the 21st annual International ACM SIGIR Conference on Research and development in information retrieval, pp. 104–111. ACM Press, New York (1998)

15. Boldi, P., Lonati, V., Santini, M., Vigna, S.: Graph fibrations, graph isomorphism, and PageRank. RAIRO Inform. Théor. 40, 227–253 (2006)

16. Boldi, P., Posenato, R., Santini, M., Vigna, S.: Traps and pitfalls of topic-biased pagerank. In: Aiello, W., Broder, A., Janssen, J., Milios, E.E. (eds.) WAW 2006. LNCS, vol. 4936, pp. 107–116. Springer, Heidelberg (2008) (Revised Papers)

17. Boldi, P., Vigna, S.: The webgraph framework I: compression techniques. In: WWW 2004: Proceedings of the 13th International Conference on World Wide Web, pp. 595–602. ACM, New York (2004), http://doi.acm.org/10.1145/988672.988752

18. Boldi, P., Vigna, S.: Codes for the World–Wide Web. Internet Math 2(4), 405–427 (2005)

19. Bollobás, B., Riordan, O.: The diameter of a scale-free random graph. Combinatorica 24(1), 5–34 (2004), http://dx.doi.org/10.1007/s00493-004-0002-2

20. Bollobás, B., Riordan, O., Spencer, J., Tusnády, G.: The degree sequence of a scale-free random graph process. Random Struct. Algorithms 18(3), 279–290 (2001), http://dx.doi.org/10.1002/rsa.1009

21. Bonato, A.: A survey of models of the web graph. In: López-Ortiz, A., Hamel, A.M. (eds.) CAAN 2004. LNCS, vol. 3405, pp. 159–172. Springer, Heidelberg (2004)

22. Bordino, I., Boldi, P., Donato, D., Santini, M., Vigna, S.: Temporal evolution of the UK web. In: ICDM Workshops, pp. 909–918. IEEE Computer Society, Los Alamitos (2008)

23. Broder, A., Kumar, R., Maghoul, F., Raghavan, P., Rajagopalan, S., Stata, R., Tomkins, A., Wiener, J.: Graph structure in the web. Comput. Netw. 33(1-6), 309–320 (2000), http://dx.doi.org/10.1016/S1389-1286(00)00083-9

24. Broder, A.Z., Charikar, M., Frieze, A.M., Mitzenmacher, M.: Min-wise independent permutations (extended abstract). In: STOC 1998: Proceedings of the thirtieth annual ACM symposium on Theory of computing, pp. 327–336. ACM, New York (1998), http://doi.acm.org/10.1145/276698.276781

25. Buriol, L.S., Frahling, G., Leonardi, S., Marchetti-Spaccamela, A., Sohler, C.: Counting triangles in data streams. In: PODS, pp. 253–262 (2006)

26. Bush, V.: As we may think. The Atlantic Monthly 176(1), 101–108 (1945)

27. Caminero, R.C., Zavarsky, P., Mikami, Y.: Status of the African web. In: WWW 2006: Proceedings of the 15th International Conference on World Wide Web, pp. 869–870. ACM, New York (2006), http://doi.acm.org/10.1145/1135777.1135919

28. Chierichetti, F., Kumar, R., Lattanzi, S., Mitzenmacher, M., Panconesi, A., Raghavan, P.: On compressing social networks. In: KDD 2009: Proceedings of the 15th ACM SIGKDD International Conference on Knowledge discovery and data mining, pp. 219–228. ACM, New York (2009), http://doi.acm.org/10.1145/1557019.1557049

29. Cho, J., Garcia-Molina, H.: Estimating frequency of change. ACM Transactions on Internet Technology 3, 2003 (2000)

30. Demetrescu, C., Finocchi, I., Ribichini, A.: Trading off space for passes in graph streaming problems. In: SODA, pp. 714–723 (2006)

31. Donato, D., Laura, L., Leonardi, S., Millozzi, S.: The web as a graph: How far we are. ACM Trans. Internet Technol. 7(1), 4 (2007)

32. Donato, D., Leonardi, S., Millozzi, S., Tsaparas, P.: Mining the inner structure of the web graph. In: Eigth International workshop on the Web and databases WebDB, Baltimore, USA (2005)

33. Dorogovtsev, S.N., Mendes, J.F.F.: Evolution of Networks: From Biological Nets to the Internet and WWW (Physics). Oxford University Press, Inc., New York (2003)

34. Eiron, N., McCurley, K.S., Tomlin, J.A.: Ranking the web frontier. In: WWW 2004: Proceedings of the 13th International Conference on World Wide Web, pp. 309–318. ACM, New York (2004), http://doi.acm.org/10.1145/988672.988714

35. Elias, P.: Efficient storage and retrieval by content and address of static files. J. ACM 21(2), 246–260 (1974), http://doi.acm.org/10.1145/321812.321820

36. Elias, P.: Universal codeword sets and representations of the integers. IEEE Transactions on Information Theory 21(2), 194–203 (1975)

37. Erdös, P., Rényi, A.: On random graphs, I. Publicationes Mathematicae (Debrecen) 6, 290–297 (1959)

38. Fano, R.M.: On the number of bits required to implement an associative memory. Memorandum 61, Computer Structures Group, Project MAC (1971)

39. Feigenbaum, J., Kannan, S., McGregor, A., Suri, S., Zhang, J.: On graph problems in a semi-streaming model. Theor. Comput. Sci. 348(2), 207–216 (2005), http://dx.doi.org/10.1016/j.tcs.2005.09.013

40. Fielding, R., Gettys, J., Mogul, J., Frystyk, H., Masinter, L., Leach, P., Berners-Lee, T.: Hypertext transfer protocol – http/1.1 (1999)

41. Fielding, R., Gettys, J., Mogul, J., Frystyk, H., Masinter, L., Leach, P., Berners-Lee, T.: Hypertext Transfer Protocol – HTTP/1.1. RFC 2616, Draft Standard (1999), http://www.ietf.org/rfc/rfc2616.txt (Updated by RFC 2817)

42. Flake, G.W., Lawrence, S., Giles, C.L., Coetzee, F.: Self-organization and identification of web communities. IEEE Computer 35(3), 66–71 (2002)

43. Adobe Flash, http://www.adobe.com/

44. Fogaras, D., Rácz, B., Csalogány, K., Sarlós, T.: Towards scaling fully personalized pagerank: Algorithms, lower bounds, and experiments. Internet Mathematics 2(3) (2005)

45. Garrett, J.J.: Ajax: A new approach to web applications (2005), http://adaptivepath.com/ideas/essays/archives/000385.php

46. Gilbert, E.N.: Random graphs. The Annals of Mathematical Statistics 30(4), 1141–1144 (1959)

47. Gyöngyi, Z., Garcia-Molina, H., Pedersen, J.: Combating web spam with TrustRank. In: Proceedings of the 30th International Conference on Very Large Databases, pp. 576–587. Morgan Kaufmann, San Francisco (2004)

48. Haveliwala, T.H.: Topic-sensitive pagerank: a context-sensitive ranking algorithm for web search. IEEE Transactions on Knowledge and Data Engineering 15(4), 784–796 (2003)

49. Henzinger, M.R.: Hyperlink analysis for the web. IEEE Internet Computing 5(1), 45–50 (2001)

50. Henzinger, M.R., Raghavan, P., Rajagopalan, S.: Computing on data streams. In: External memory algorithms, pp. 107–118. American Mathematical Society, Boston (1999)

51. Jaccard, P.: Étude comparative de la distribution florale dans une portion des alpes et des jura. Bulletin del la Société Vaudoise des Sciences Naturelles 37, 547–579 (1901)
52. Joo, W.K., Myaeng, S.H.: Improving retrieval effectiveness with hyperlink information. In: Proceedings of International Workshop on Information Retrieval with Asian Languages (IRAL), Singapore (1998)
53. Kalyanasundaram, B., Schintger, G.: The probabilistic communication complexity of set intersection. SIAM J. Discret. Math. 5(4), 545–557 (1992), http://dx.doi.org/10.1137/0405044
54. Kamvar, S., Haveliwala, T., Golub, G.: Adaptive methods for the computation of pagerank. Technical Report 2003-26, Stanford InfoLab (2003)
55. Kamvar, S.D., Haveliwala, T.H., Manning, C.D., Golub, G.H.: Extrapolation methods for accelerating pagerank computations. In: WWW 2003: Proceedings of the 12th International Conference on World Wide Web, pp. 261–270. ACM, New York (2003), http://doi.acm.org/10.1145/775152.775190
56. Kernighan, B.W., Lin, S.: An efficient heuristic procedure for partitioning graphs. The Bell system technical journal 49(1), 291–307 (1970)
57. Kleinberg, J.M.: Authoritative sources in a hyperlinked environment. J. ACM 46(5), 604–632 (1999), http://doi.acm.org/10.1145/324133.324140
58. Langville, A.N., Meyer, C.D.: Deeper inside pagerank. Internet Mathematics 1 (2004)
59. Lee, C.P.C., Golub, G., Zenios, S.A.: Partial state space aggregation based on lumpability and its application to PageRank. Technical report, Stanford InfoLab (2003)
60. Lempel, R., Moran, S.: The stochastic approach for link-structure analysis (SALSA) and the TKC effect. Comput. Netw. 33(1-6), 387–401 (2000), http://dx.doi.org/10.1016/S1389-12860000034-7
61. Levenstein, V.E.: On the redundancy and delay of separable codes for the natural numbers. Problems of Cybernetics 20, 173–179 (1968)
62. Li, Y.: Toward a qualitative search engine. IEEE Internet Computing 2(4), 24–29 (1998)
63. Marchiori, M.: The quest for correct information of the Web: hyper search engines. In: Proc. of the sixth international conference on the Web, Santa Clara, CA, USA, pp. 265–274 (1997)
64. Najork, M.A., Zaragoza, H., Taylor, M.J.: HITS on the web: how does it compare? In: SIGIR 2007: Proceedings of the 30th annual International ACM SIGIR Conference on Research and development in information retrieval, pp. 471–478. ACM, New York (2007), http://doi.acm.org/10.1145/1277741.1277823
65. Ntoulas, A., Cho, J., Olston, C.: What's new on the web?: the evolution of the web from a search engine perspective. In: WWW 2004: Proceedings of the 13th International Conference on World Wide Web, pp. 1–12. ACM, New York (2004), http://doi.acm.org/10.1145/988672.988674
66. Page, L., Brin, S., Motwani, R., Inograd, T.: The pagerank citation ranking: Bringing order to the web. Technical Report 1999-66, Stanford InfoLab (1999), Previous number = SIDL-WP-1999-0120
67. Pinski, G., Narin, F.: Citation influence for journal aggregates of scientific publications: Theory, with application to the literature of Physics. IP&M 12, 297–326 (1976)
68. Porter, M.A., Onnela, J.P., Mucha, P.J.: Communities in networks. Notices of the American Mathematical Society 56(9), 1082–1097 (2009)
69. Raghavan, S., Garcia-Molina, H.: Representing web graphs. Technical Report 2002-30, Stanford InfoLab (2002)
70. Randall, K., Stata, R., Wickremesinghe, R., Wiener, J.L.: The LINK database: Fast access to graphs of the Web. Research Report 175, Compaq Systems Research Center, Palo Alto, CA (2001)

71. Salomon, D.: Variable-length Codes for Data Compression. Springer-Verlag New York, Inc., Secaucus (2007)
72. Suel, T., Yuan, J.: Compressing the graph structure of the web. In: Data Compression Conference, vol. 0, p. 0213 (2001),
    http://doi.ieeecomputersociety.org/10.1109/DCC.2001.917152
73. Upstill, T.: Predicting fame and fortune: Pagerank or indegree? In: Proceedings of the Australasian Document Computing Symposium, ADCS 2003, pp. 31–40 (2003)
74. W3C: Deprecated usage of the META element for redirects,
    http://www.w3.org/TR/WCAG10-HTML-TECHS/#meta-element
75. Wasserman, S., Faust, K., Iacobucci, D.: Social Network Analysis: Methods and Applications (Structural Analysis in the Social Sciences). Cambridge University Press, Cambridge (1994)
76. Watts, D.J., Strogatz, S.H.: Collective dynamics of small-world networks. Nature 393(6684), 440–442 (1998)
77. Wu, B., Davison, B.D.: Identifying link farm spam pages. In: Proceedings of the 14th International World Wide Web Conference, Industrial Track (2005)
78. Zhu, J.J.H., Meng, T., Xie, Z., Li, G., Li, X.: A teapot graph and its hierarchical structure of the Chinese web. In: WWW 2008: Proceeding of the 17th International Conference on World Wide Web, pp. 1133–1134. ACM, New York (2008),
    http://doi.acm.org/10.1145/1367497.1367692

Chapter 6
Web Usage Mining

Pablo E. Román*, Gastón L'Huillier, and Juan D. Velásquez

Abstract. In recent years, e-businesses have been profiting from recent advances on the analysis of web customer behaviour. For decades experts have debated on ways of presenting the content or structure in a web site in order to captivate the attention of the web user in the web intelligence community. A solution to this could help boost sales in an e-commerce site. Web Usage Mining (WUM) is the extraction of the web user browsing behaviour using data mining techniques on web data. According to this, several models of data analysis have been used to characterize the Web User Browsing Behaviour. Nevertheless, outstanding techniques have recently developed in order to improve the conventional success rates for behavioural pattern extraction. In this chapter different approaches for WUM are presented, considering their main insights, results, and applications to web behaviour systems.

6.1 Introduction

The Internet has become a regular channel for communication, most of all for business transactions. Commerce over the Internet has grown to higher levels in recent years. For instance, e-shopping sales has been drastically increasing in the previous year, achieving a growth of 17% in 2007,generating revenue of $240 billion/Year in the US alone [21]. This highlights the importance of acquiring the knowledge on how the Internet monitors customer's interaction within a particular web site.

One can compare this new technological environment using traditional marketing approaches, but the internet embraces new methods of determining consumers genuine needs and tastes. Traditional market surveys serve no purpose in reflecting the veracious requirements of customers who have not been precisely defined in the web

Pablo Román · Gastón L'Huillier · Juan D. Velásquez
Department of Industrial Engineering, University of Chile, República 701, Santiago, Chile
e-mail: `proman@ing.uchile.cl`, `glhuilli@dcc.uchile.cl`,
`jvelasqu@dii.uchile.cl`

* Corresponding author.

J.D. Velásquez and L.C. Jain (Eds.): Advanced Techniques in Web Intelligence – 1, SCI 311, pp. 143–165.
springerlink.com   © Springer-Verlag Berlin Heidelberg 2010

context. It is well known that Web users are ubiquitous. In this sense, a marketing survey compiled on a specific location in the world does not carry clear statistical significance. However, online queries should improve this issue by requesting that each visitor answers as many focused question as they can [37]], but apart from predicting future customer preferences, online surveys can improve the effectiveness of the web site content strategy.

Web Usage Mining (WUM) can be defined as the application of machine learning techniques over web data for automatic extraction of behavioural patterns from web users . In this sense, the web usage patterns can be used for analyzing the web user preferences. Traditional data mining methods need to be pre-processed and adapted before employing over web data. Several efforts have been made to improve the quality of the resulting data, which are described in the data pre-processing chapter of this book. Once a repository of web user behaviour (Web Warehouse) is available [83], specific machine learning algorithms are applied in order to extract pattern regarding the usage of the web site. As a result of this process several applications can be implemented as adaptive web sites, such as recommender systems, and revenue management marketing amongst others.

For instance, the problems connected to the customization of web sites that are geared to improve sales are somewhat challenging. Recently, the one million dollar Netflix prize [46] has been contested after three years without a winner. NetFlix is an online DVD rental company, and its business is driven by online movie recommendations for customers based on its ratings of movie. The competition consists of improving the forecasting algorithm of the company for rating per users. The 2009 winning team used a modified linear stacked generalization algorithm for recommendations [89] only for improving the performance of the Netflix predictive algorithm by 10 percent. This is an example of how difficult it is to track or monitor the web user behaviour, but also it highlights the real life importance of such predictions. In this chapter the state of the art of WUM coupled with new trends that face the discipline is discussed.

6.2 Characterizing the Web User Browsing Behaviour

As described in [34, 49, 76, 78, 83], Web usage data can be extracted from different sources, from which web logs are considered as one of the main resources for web mining applications. The variety of different sources carries a number of complexities in terms of data pre-processing and furthermore these are associated with the incompleteness of each source. As a solution to this several pre-processing algorithms have been developed [83]. Further sources like the hyperlink structure and the web content complement the extracted Log information, providing a semantic dimension of each user's action. Furthermore, in terms of the web log entries, several problems must be confronted. Overall, a web log in itself does not necessarily reflect a sequence of an individual user's documented access. Instead it registers every retrieval action but without a unique identification for each user.

6.2.1 Representative Variables

The web user browsing behaviour can be monitored by three kinds of data: the web structure, the web content and the web user session. The first is directly related to the environment. The third describes the click stream that each Web User performs during its visit to the web site.

- **The web structure:** A Web Site can be represented as a directed graph $G(N,V,T)$, consisting of a collection of n nodes $N = \{1,\ldots,n\}$ and vertices $V = \{(i,j)/\ a\ web\ link\ point\ from\ i\ to\ j\}$ with text content $T = \{T_i\}$. A node i from G corresponds to a web page with text content T_i, the representation of the content will be described later. Two special nodes need to be individualized, as they do not correspond to any real page. This is because they represent the exit/entrance to the web site and each node consists of a link to the "exit or entrance" node. This representation has the advantage of explicitly including all transitions between nodes, which is useful for stochastic process descriptions.

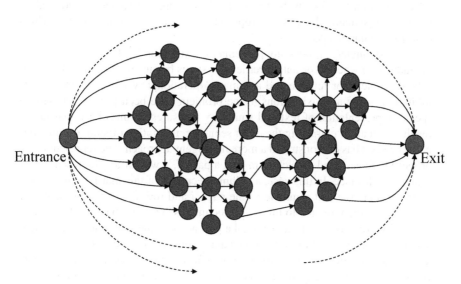

Fig. 6.1 Static graph web site structure representation

Nevertheless, this description of a Web Site can be considered as a first approximation of the real hyperlink structure. The notion of a web page consisting of static content and unique URL can not fit the dynamical case. Web sites are being continuously updated, persistently modifying links and content, including those that depend on web user profile adaptive Web Site changes . On static pages with frames this concept is also challenged since the page is a composite. As stated in the Internet Report of China [22], the rate of web pages that moves from static to dynamic content is close to one.

Considering web 2.0 sites , the latter model seems obsolete. However, this approximation is valid considering some general circumstances. Informative sites that are updated on a regular periodic basis like those of newspapers can be represented under this graph representation within a definite period of time. Blogs can also be represented as such considering that the replies are increasingly added due to the fact that their nodes consist of post and reply.

More generally a time dependent graph structure of a variety of web objects and its associations can be defined for general purpose sites. For application, the analysis should be adjusted to the simpler model depending on the particular web site. Other representations of the structure of the web site are discussed in the data pre-processing Chapter.

- **The web content:** This corresponds to the web user perception semantic information of each visited page from a Web Site. On the earlier Internet this data corresponded mainly to the text content. Nowadays, Web 2.0 sites represent a much more complex picture. The web content is more dynamic and is constituted by a rich variety of media (text, images, video, embedded application, etc.). Web pages are composites of web objects that have semantic values. Several valuations of the semantic have been proposed and are revised in the chapter which describes semantic issues.

  Natural language processing for semantic extraction has been a large subject of study from when information retrieval systems began and it is still a large unsolved problem for reliable and automatic operational system [48]. Despite its limitation, some approximations to the problem have been proposed based on special representation of text and similarity measures at times have reasonable results. Instead of extracting the exact semantic, the notion of similar semantics between texts has demonstrated a more intelligent approach. The representations proposed are assigned to each term p in a page j, a weight ω_{pj} representing its semantic relevance. In this case, a column vector d_j of this matrix ω represents approximately the text semantic of the page j and row vector $t_p^{\dagger}$ represents the term p semanticly related to the set of documents. This approach is also termed "Bag of Word" because it does not take into account the semantic relations and syntax of phrases. In this way every equal word has the same importance independently of the context. Similar approaches can be extended to non-textual content like web object, where meta-data plays a fundamental role in representing the semantic.

  Furthermore, dynamic content implies time and user dependence of the semantic. As web applications become more complex than the standard representation of the content the semantics become more inaccurate. Specific semantic context that group a variety of content must be tailored for each specific application.

- **The web user session:** Web User visits a Web Site represented by the browsing trajectorie that is categorised as a session [74]. A session s is a sequence $s = [(i_1, \tau_1), \ldots, (i_L, \tau_L)] \in S$ of pages $i_k \in N$ and time $\tau_k \in \mathbb{R}^+$ spend by a Web User. The size $L = \|s\|$ of a session corresponds to the number of nodes without considering the sink and source. In this representation the time associated with both the source and sink nodes and the duration of a session $\mathscr{T} = \sum_k \tau_k$ is the sum of all visitors times spent on the site.

Nevertheless, if sessions are not explicitly given they must be reconstructed from other sources like web logs. When they are specially retrieved some privacy concerns [50, 36] arise that complicate its implementation. On the other hand, session retrieval from web logs have less relation to privacy issues since the data is stored anonymously. The process of extracting sessions has been reviewed in the chapter entitled data pre-processing.

Nevertheless, web data has some concerns. However, a further problem is associated with this data: the high dimensionality. Data mining algorithms suffer from the so called "curse of dimensionality" phenomenon. Over the years the processing of such data has been specialized as the Web Mining discipline, in particular when the purpose of such analysis is related to the behaviour of the Web User. In such cases it is called Web Usage Mining.

Also some evident problems exist that are connected with the web usage data. For example, the high diversity of some web pages; search engines that allow users to directly access some part of the web site; a single IP address with single server sessions; single IP address with multiple server sessions; multiple IP address with single server sessions; multiple IP address associated with a single visitor; and multiple agents associated with a single user session [83]. Additionally, a user's activation of the forward and reverse browser button is often not recorded in the web log because, in most cases, the browser retrieves the page from its own cache. A proxy server, acting as an internet web page cache serves to reduce network traffic, and can also capture web requests that are not recorded in a web log [20].

Browsing data has been recently considered in WUM, where the scroll-bar, select and save-as user interactions with the web site [78, 79]. Furthermore, semantic considerations have been proposed by different authors, where Plumbaum et al. in [61] uses the open standard of Microformats in order to add semantic information on the web page. This is a similar method as proposed in [70], for which JavaScript events (gathered with AJAX) are associated with key concepts in the portal, providing a context of such events and linking valuable usage information with the semantic of the web site.

As stated in [26], WUM presents different challenging problems in terms of the pre-processing of the usage data. Weblogs are larger in size, and both data volume and dimensionality, where sparse dataset representations of web users are needed to be transformed into a more accurate user behaviour representation. One of the problems associated to this lies in the high dimensionality, which results in computational complexity of the mining process. Secondly, the data scarcity results in the mining algorithms having the ability to extract meaningful and interesting patterns in the user browsing behaviour.

Each WUM technique requires a model of user behaviour per web site in order to define a feature vector to extract behaviour patterns. Usually, the model contains the sequence of pages visited during the user session and some usage statistics, like the time spent per session, and pages viewed, amongst other information gathered. Here difficulties can be encountered when a page is loaded into a given web browser, the request for the web site objects can be logged separately, for which a series of page can be viewed associated with the same session.

All of the latter leads to the fact that web usage data requires different pre-processing techniques before analyzing the user behaviour [76]. However, one of the main tasks present in WUM is the determination of the web user sessions based on the web usage data collated from a given web site (sessionization). It is well known that strategies for sessionization can be classified as reactive and proactive [30].

Proactive sessionization strategies capture a rich collection of a web user's activity during the visit to a given web site. However, this practice is considered invasive, and even forbidden in some countries [74], or regulated by law to protect the user's privacy [43]. Examples of these methods include cookie oriented session retrieval [4], URL rewriting [18], and web tracking software, close to spyware, installed on the user's computer (or browser) to capture the entire session [50].

Reactive sessionization strategies have less privacy concerns because they are design to use only the web log entries' information, which excludes explicit user information [74]. However, a web log only provides an approximate way of retrieving a user's session for previously stated reasons. This reinforces the need to reconstruct a user's session from the information available (sessionization). Prior work on sessionization has relied on heuristics [3, 12, 74] which have been applied with a high degree of success on a variety of studies that include web user navigational behaviour, recommender systems, pattern extraction, and web site keyword analysis [83].

6.2.2   Empirical Statistics Studies about Web Usage

The human behaviour shows some predictive regularity on the averages, but to the contrary of the free will hypothesis. Some of those regularities are observed on the distributions of session in different kinds of web sites [29]. With the help of such regular statistics of the human behaviour Web Usage Mining can be tailored to fit those conditions. Several stochastic models have been theorized in order to mathematically explain this result [29, 80, 81], but nothing is related intrinsically to the physical phenomena. Some others models based on the neurophysiology of the decision making process have also been proposed [66].

Data mining processing on web data should show results that are in agreement with the observed universal probability distributions. As it was commented, web user's actions on a web site follow regular patterns in probability distribution. This is additional information that can reduce the size of the feature space for machine learning algorithm. When such kind of reduction is available, algorithms have a narrow region for working resulting in better performance and accuracy. Nevertheless procedures must be adapted for fitting such statistical constrains [45, 57]. Understanding those statistics results in a better standard and quality for user [87].

Some important statistical empirical studies are summarized below.

- **Session Length Distribution:** Empirical studies over different web sites shows that the distribution of session size follows a common shaped function having an asymptotic heavy tail. Following [29] an Inverse Gaussian distribution ties in well with reality, and it was termed the universal law of surfing. In some work a Zipf distribution (power law) has been observed [44] reflecting a real session, but this distribution is used to approximate the Inverse Gaussian since its tails decay much slower than a Gaussian. Application of this kind of regularities enables the tuning up of systems like web crawler [2] and session retrieval from log file [13]. This web usage regularity has also been exploited for mining [45].

- **Information seeking behaviour:** While studies focus on algorithm for pattern extraction, few research relate to how web task users perform. Furthermore, the manner in which people seek information through the web can be classified through cognitive styles [88]. The studies differentiated two kind of web user: Navigators (17% of total of users) and Explorers (3%). The first, maintain consistency on the sequence of pages of visited pages. Navigators seek information sequentially and revisit the same sites frequently. The second have highly variable pattern of navigation. Explorers have a tendency to query web search pages frequently, revisit pages several times and browse a large variety of web sites. This kind of statistical study highlights the impact of classical navigational pattern extraction. Navigator will have the most influence on data regardless of consisting of 17 percent of total use. Nonetheless a whole skew distribution of cognitive styles have been reported [88], beginning with Navigators and ending with Explorers web users. The study of this distribution needs to be taken in account for further specialization of usage mining studies. A large study of cognitive information seeking has been investigated [32] showing that context can influence each particular web user behaviours. Others sources focus [38] on the statistics of the task that web users perform. Such taxonomy becomes associated with web user who perform transactions (e.g. reading emails 46.7%), Browsing (e.g. news reading 19.9%), Fact Finding (e.g. looking for whether 16.3%), Info Gathering (e.g. job hunting 13.5%) and a 1.7% non-classified. These four categories specify a simpler structure concerning the web usage.
- **Web User Habits:** Web user's habits have changed since the internet became more sophisticated. Nevertheless with current new web 2.0 applications web logs are becoming a much more intricate data source. Web users browsing behaviour are for ever changing, using fewer backtracking support tools such as the back button. However there is an increasing usage of parallel browsing with tabs and new windows.
- **The inter-event time for web site visit:** Similar heavy tailed distribution [56] has been measured on the time spent on pages throughout wide variety of the pages.

6.2.3   Amateur and Expert Users

Users can be grouped into two categories: experienced and inexperienced or "amateurs" [83]. The latter is unfamiliar with the process of accessing web sites and possibly dealing with web technology. Their behaviour is characterized by erratic browsing and sometimes they do not find what they are looking for. The former are users with web site experience and with some standard knowledge of web technology. Their behaviour is characterized by spending little time on pages with low interest and thus concentrating on the pages they are looking for, on which where they spend a significant amount of time. As amateurs gain experience, they slowly become experienced users who are aware of a particular web site's features. Therefore recommendations for change should be based on those users.

On the one hand, amateur users correspond to those unfamiliar with a particular web site and most probably with web technology skills [83]. Their browsing behaviour is erratic and often they do not find what they are looking for. On the other hand, experienced users are familiar with this or similar sites and have a certain degree of web technology skills. They tend to spend little time visiting low interest pages and concentrate on the pages they are looking for on which they spend a significant amount of time. As amateurs gain skills they slowly become experienced users, and spend more time on pages that interest them.

6.3   Representing the Web User Browsing Behaviour and Preferences

Regarding data mining purposes there are two kinds of structures that are used: features vectors that correspond to tuples of real number, and graph representation where numeric attributes are associated with nodes and relations. The most used representation corresponds to feature vectors with information about sessions, web page content and web site structure. Feature vector are employed for traditional web mining in an unsupervised a supervised fashion. Furthermore, Graph data structure is used for graph mining techniques or rule extraction.

6.3.1   Vector Representations

A session is a variable length of data structure that is not directly usable for most of the data mining algorithm. Web user activities can be extracted in several ways for the summarization of a session which is codified by mean of weighting ω_i the usage per page i. The vector $v = [\omega_i] \in V^n \subset \mathbb{R}^n$ has a dimension n corresponding to the number of different web pages. The cardinality of the set $|V| = m$ is equal to the number of collected session in the pre-processing phase. One of the most important information that this representation does not reflect, it is the sequence logic of each session. Despite this simplification this methodology has been used with success in several web mining applications.

Several methods exist to evaluate the weight ω_i. The simplest weighting schema corresponds to assigning a binary value $\omega_i \in \{0, 1\}$ represented if the page is used (1) or not (0) on this session [54]. More information can be incorporated extending the binary passage of the web page by the visit duration fraction. In this case the weight remains in the interval $\omega \in [0, 1]$ for normalization purposes. The fraction of time remaining in a web page is supposed to be an indicator of the quality and interest of the content [55]. Other weighting measures attempt at using other prior information about pages for measuring the significance of each page [55].

6.3.2   Incorporating Content Valuations

Hypermedia can be measured primarily by its text content. Natural Language Processing techniques consist of measuring text by means of different multidimensional representation. The most common valuation is the vector space model, where a document i is represented by a vector $m_i = [m_{ij}]$. Each component represents a weight for the importance of the word j in this document i. This model uses the "Bag of Word" abstraction, where any sentence structure is disregarded in favour of simple word frequencies. Furthermore, this semantic approximation has demostrated accurate results for data mining application. Several weighting schemas have been used with different results. The simplest is the binary weighting scheme where $m_{ij} = 1$ if the term j is present on the document i. The most used weighting scheme is the TF-IDF that combines the frequency of the term j in the document i and the frequency of document containing the term i. Recently the weight has been constructed with the help of a machine learning feature selection mechanism [42].

The text weighting scheme m_{ij} enriches the information provided by the feature vector of web usage. The visitor behaviour vector $v = [(m_i, \omega_i)]_{i=1}^n$, which each component represent the content and page importance.

6.3.3   Web Object Valuation

Nowadays web sites become highly dynamic applications. The visual presentation of a web page on a browser can not be identified correctly with a URL. A variable multiplicity of hypermedia could appear on browser presentation for the same URL. Nevertheless, embedded visual entities on web pages seem to be a more reliable concept. An object displayed within a webpage is termed a Web Object. Despite the complex semantic analysis of multimedia, metadata is used to define the Web Object within it (Figure 6.2).

Meta data that describes web objects constitute the information source for building the vector representation of content. The user's point of view is the principal research topic from which Web Object techniques has been developed. In this way the content and appearance of a web page is combined for processing. Different ways have been developed to describe web pages based on how the user perceives a particular page.

Web Object research has been carried in the following work: Web site Key objects identification [16], Web Page Element Classification [8], Named Objects [73] and Entity extraction from the Web.

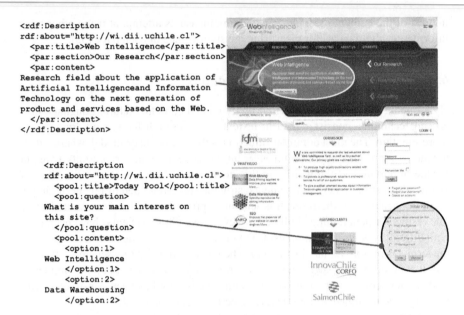

```
<rdf:Description
rdf:about="http://wi.dii.uchile.cl">
  <par:title>Web Intelligence</par:title>
  <par:section>Our Research</par:section>
  <par:content>
Research field about the application of
Artificial Intelligenceand Information
Technology on the next generation of
product and services based on the Web.
  </par:content>
</rdf:Description>
```

```
<rdf:Description
rdf:about="http://wi.dii.uchile.cl">
  <pool:title>Today Pool</pool:title>
  <pool:question>
What is your main interest on
this site?
  </pool:question>
  <pool:content>
    <option:1>
Web Intelligence
    </option:1>
    <option:2>
Data Warehousing
    </option:2>
```

Fig. 6.2 Some Web Object identified visually on browser and its XML representation

6.3.4 Graph Representation

Graph mining uses graph theoretical construct and algorithm for discovering patterns in data [10]. On [23] web user trails are converted into a weighted graph using a similarity measure between session. Nodes correspond to sessions and arcs are labelled by the value of the similarity measure between both nodes. This similarity measure is used to overlap degrees between sessions, but any other measure can be adopted. The resulting structure is a direct representation of web user behaviour similarity.

6.3.5 The High Dimensionality of Representation

Usage data corresponds to a high dimensional representation. Considering that a medium sized web site contains thousand of pages and that around ten thousand terms exist, therefore the feature vector dimensionality correspond to at least 10^4 components. Automatic data mining methods based on similarity measure suffers from the "curse of dimensionality". This problem corresponds to the exponential grows of the size of the search space for data mining algorithm. Performance issues are one of the principal problems that could render the problem intractable. Recent studies reveal that distance based algorithm is affected since numeric difference between distance measure collapses in higher dimensional space. Similarity measure based algorithms (e.g. Clustering) are highly affected by this phenomenon.

Feature selection is a technique for data refinement that has been recently used for alleviating the higher dimensionality problem [31, 90]. A radical method of reducing the dimension of the feature vector is by way of a supervised approach to the text representation of web pages [63]. In this case an expert categorizes the semantic of the web pages which help to reduce the dimension of the feature vector.

6.4   Extracting Patterns from Web User Browsing Behaviour

As defined by Srivastava et al. in [76], "Web usage mining (WUM) is the application of data mining techniques to discover usage patterns from Web data, in order to understand the needs of Web-based application". Given this, one of the most challenging topics is the understanding and discovery of usage patterns from human users. Moreover, to analyze and predict human behaviour is its main characteristic, which is differentiated from Web Structure Mining and Web Content Mining, where techniques adopted are from a different nature and span between researchers and practitioners. It is relevant to point out that WUM is sometimes referred to as click stream analysis [76], considered as the analysis and extraction of underlying patterns from an aggregated sequence of page visits from a particular web site user navigation.

Interest in WUM is growing rapidly in the scientific and commercial communities, possibly due to both its direct application to web personalization and the increased complexity of web sites [83], and Web 2.0 applications [61]. In general, WUM uses traditional behavioural models, operations research and data mining methods (which will be intensively reviewed in this chapter) that deal with web usage data. However, some modifications are necessary according to their respective application domain, due to the different types of web usage data.

In general terms, two families of techniques have been used to analyze sequential patterns: deterministic and stochastic techniques. Each one of these techniques gathers different optimization methods, from data mining, operations research, and stochastic modelling, where different approaches have been adopted to compensate for the lack of analysis regarding some of these techniques present.

6.4.1   Clustering Analysis

Clustering user sessions can be used, essentially, for grouping users with common browsing behaviour and those interested in pages with similar content [76]. In both cases, the application of clustering techniques is straightforward when extracting patterns to improve the web site structure and content. Normally, these improvements are carried out at site shut-down.

However, there are systems that personalize user navigation with online recommendations about which page or content should be visited [60, 84]. In [35, 67], the *k*-means k-means clustering method is used to create navigation clusters. In [91], the usage vector is represented by the set of pages visited during the session, where the similarity measure considers the set of common pages visited during the user session.

Following previously stated reasoning, in [35], the pages visited are also considered for similarity with reference to the structure of the web site as well as the URLs involved. In [27] and [67], the sequence of visited pages is incorporated into the similarity measures additional to the page usage data. In [82], it is proposed that, despite the page sequence, the time spent per page during the session and the text content in each page is included in the similarity measure.

Also, clustering techniques such as Self-Organizing Features Maps (SOFM) have been used to group user sessions and information that could lead to the correct characterization of the web user behaviour [83]. In this context, Velásquez et al. in [85] proposed two feature vectors, composed by the information related to the text content from a web site, and a feature vector related to the usage, including information such as the time spend on the visited page by the web user. With this, the authors obtained the relevant words from the web site (or keywords), methodology which was extended by Dujovne et al. in [16] to determine the web site key objects.

Graph clustering techniques are applied to graph web data representation; through which [23] the time spent on each page was considered for similarity calculation and graph construction. The representation of a session is created by using a similarity measure on the sequence of visited pages. A weighted graph is constructed using a similarity measure, and this in turn is achieved by employing an algorithm of sequence alignment [23]. This representation is further processed using a graph clustering algorithm for web user classification. Association rules can also be extracted [14].

Recently, Park et al. in [58], refers to the relevance of clustering in WUM that aims to find groups whose common interests and behaviour are shared. In this work, the question is whether sequence based clustering performed more effectively than frequency based, and thus acquired the best results a sequence based fuzzy clustering methodology proposed by the authors. Likewise, another recent clustering based methodology in WUM was proposed by Rios et al. in [63], where a semantic analysis was developed by considering a concept-based approach for off-line Web site enhancements. Here, the main development was the introduction of concepts into the mining process of WUM, carried out by a hybrid method based on the evaluation of web sites using Nakanishi's fuzzy reasoning model, and a similarity measures for semantic web usage mining.

6.4.2   Decision Rules

Decision rule induction is one of the classification approaches widely used in web usage mining [76]. Its practical results are sets of rules that represent the users' interests. In WUM, the association rules are focused mainly on the discovery of relations between the pages visited by the users [52]. For instance, an association rule for a Master and Business Administration (MBA) program is mba/seminar.html mba/speakers.html, showing that a user who is interested in a seminar tends to visit the

speaker information page. Based on the extraction rules, it is possible to personalize
web site information for a particular user [53].

6.4.3   Integer Programming

On terms of deterministic sessionization, Dell et al. in [13] presents a session reconstruction algorithm based on integer programming. This approach presents different advantages to address linking structure constraints presented by the time sequence of the web log entries, finding the best combinations of path that fulfill these
constraints. Recent advances on integer programming and optimization [5], allows
solving hard combinatorial problem with an acceptable timing. This method will be
presented with further details in the data pre-processing chapter.

6.4.4   Markov Chain Models

Several statistical models for web surfing and web user behaviour have been developed in [6, 11, 17, 33, 75]. Here Markov models have been proposed for the
modelling of the behaviour of the web user. In this cases, a web site is considered
as an undirected graph of pages, in which the web user passes from one page to
another [33, 64] within an estimated probability.

On the basis of a large number of users, and taking into consideration that the web
site browsing is performed during a large period of time, it is possible to predict
transition probabilities between pages. It can be considered that Web users have
limited memory about visited pages, through which a transition probabilities can be
considered independent of older transition probabilities. If transition probabilities
are independent in more than k previous stages then the chain is known as a k-order
Markov chain. Let X_o be the page visited at the step o and then a k-order Markov
chain must have property presented in equation 6.1,

$$P(X_o|X_{o-1},\ldots,X_1) = P(X_o|X_{o-1},\ldots,X_{o-k}) \qquad (6.1)$$

The latter expression represents a stochastic process with a k-step memory. A k-
order Markov chain can be represented as a first-order Markov chain by re-labelling
techniques [62].

Once the web user behaviour is modelled, and its flow probabilities determined,
different levels of analysis and information can be extracted. Therefore, the prediction of the session size distribution, sequence of pages more likely to be visited, the
mean time that a user can spend on the site or the ranking of the page that is most
likely visited, are some of the possible outcomes that could be determined for the
decision making process in a given web site.

Usually, a web site has a large amount of web pages and building transition matrices for the Markov model could be costly. For this reason, some authors [1] recommend the reduction of the dimensionality using clustering techniques over web
pages. Then, the site transitions could be interpreted as web users changing from

different clusters. The predictive power of the Markov chain has been studied by Sarukkai in [69], where the next browsing step of a web user is constructed taking the next link as the one with maximum probability. However, some authors proposed higher order Markov models to use for this task [15, 93], considering that lower order Markov models have been found with poor predictive power [9].

6.4.5 Mixture of Markov Models

Mixtures of Markov chains, a discrete set of Markov chains that represents different web user groups with different browsing behaviour [9, 71], has been proposed as an alternative for the web usage pattern extraction with emphasis on the different types of users a site might have. In this case, the independent relationship with past behaviour is represented by an extended mixture coefficient $P(k) = \lambda_k$ into $P(X_o|X_{o-1}, \ldots, X_1) = \sum_{k=1}^{K} P(X_o|X_{o-1}, k)P(k)$, which represents K different browsing behaviour determined by further data mining techniques.

6.4.6 Hidden Markov Models

As an extension of the traditional Markov chain modelling, Hidden Markov Models (HMM) have been used to model the stochastic representation of the web site. However considering hidden states in the usage itself, web users have a underlying patterns in their behaviour [19, 72, 86]. Overall, this stochastic modelling tool has been used to determine frequent interest navigation patterns, combining web usage data and WCM techniques to build a predictive model.

6.4.7 Conditional Random Fields

Other approaches used to predict the web user behaviour are based on Conditional Random Fields (CRF) [41]. CRFs are a probabilistic framework generally used for the classification of sequential data. In the WUM context, Guo et al. in [25] aimed to predict all of the probable subsequent web pages for web users, comparing its results to other well known probabilistic frameworks, such as plain Markov chains and HMMs. Subsequently, in [24], an Error Correcting Output Coding (ECOC) of the CRF was proposed for the prediction of subsequent web pages on large-size web sites, extending previous development to a multi-class classification task for the web site prediction problem. It compared its results against single multi-label CRFs which outperformed the proposed method.

6.4.8 Variable Length Markov Chain (VLMC)

Another probabilistic framework developed for the web user behaviour modelling was proposed by Borges et al in [7], using as Variable Length Markov Chain (VLMC). There models are based on a non-fixed memory size, for which its usage in web browsing behaviour modelling is considerable. In [7], the main purpose for the researchers was to incorporate the dynamic extension of some web navigation sessions. They also aimed to extend the plain Markov chain method into a more general predictive tool. In this work, authors proved that the usage of such techniques increases the prediction accuracy, as well as the summarization ability of the Markov chain, an effective tool for the personalization of web sites, given that the web usage data is gathered by practitioners.

6.4.9   Biology Inspired Web User Model

A rather different approach from data mining consists of modelling the web user as a human operating a browser and performing decision about which hyperlink to click [66]. Neurophysiological stochastic model [65] has been studied from forty years ago establishing the stochastic evolution (Equation 6.2) of neuronal activity (Y_i) on the brain while the subject reaches a decision (Figure 6.3). Parameter of this model are (κ, λ, σ) that are related to individual neuron configuration and the most important is I_i representing the likelihood of the choice i.

$$dY_i = (-\kappa Y_i - \lambda \sum_{j \ i} f(Y_j) + I_i)dt + \sigma dW_i \qquad (6.2)$$

Applying this theory to the browsing decision process leads to a complete and effective stochastic model of a web user. The model is based on the utility preferences per hyperlink. This utility function is modelled considering how each user is related to a text preference vector μ and a text similarity function between this vector and the hyperlink's text content representation. The user utility is higher if the hyperlink content vector is more similar to μ. A random utility model generates the choice a priori probability (I_i). The similarity measure used is the cosine between vector and the weighting scheme is the TF-IDF [66].

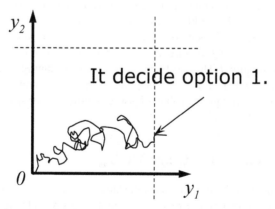

Fig. 6.3 Stochastic process for decision making. A decision is taken at the first hitting coordinate's time

This new class of model has the capability to simulate the hyperlink aggregated demand and staying time on each page. This capability depends on the parameter adjustment using the available web usage data. A sub-product of the parameter fitting process is the probability distribution for text preferences $P(\mu)$. Furthermore predicting hyperlink is based on the content and structural modifications that can be obtained using stochastic simulation.

6.4.10   Ant Colony Models

Ant colonies have been used to learn the web usage [47]. The model is based on a simplification of a model based on the neurophysiology of the human decision making [66] and the random surfer [68]. The model called the "Ant Surfer" is where agents evolve like in the original random surfer model. The Ant Surfer start in a random page and continue browsing with probability p or return to the nest with probability $(1 - p)$. The agent objective is foraging for information that is accumulated and its satiation is modelled with a threshold on the accumulated information utility. When the threshold is reach then it returns to the nest. The agent positioned on the page i select the hyperlink j to follows with probability P_{ij}, that correspond to the Logit model with utility given by the similarity measure between the ant text preference μ and the hyperlink text [66]. This model is applied to extract the web user preference and to predict web usage. Others models relate to other Markovian models for measuring the navigability of a web site [92] proposing similar web user models.

6.4.11   Matrix Factorization Methods

The NetFlix price [46] has been a corner stone evident to the magnitude of the Web Usage problem. A one million dollars price was announced for improving 10% the RMS error of the current NetFlix movie rental recommendation. The algorithm is based on matrix factorization which has been proven superior to similarity measure based techniques [40].The problem lies in mapping web usage vector and product vector (books, movies). In this model a join vector (u_l, p_k) of web user l behaviour vector u_l and the product k feature vector p_k and is linearly mapped to a factor space of dimensionality f.This problem stems from the family of singular value decomposition [59] where the linear mapping is partially known. The algorithm solves a minimum error square problem between known factor space rating values and user's and product data [77].

6.5   Application of Web Usage Mining

Web usage mining enables the analysis of the habits of a web user browsing in a web site. Furthermore knowing the user's interest and browsing behaviour can be used to improve a web site or build new web applications. The web usage mining could be used with automatic On-line algorithm, or in an Off-line fashion, supervised techniques. The general framework for the application of the web usage mining is the adaptive web sites [83]. The automatic personalization of the web site to the web users tastes, habits or marketing recommendation is only one of the techniques used on adaptive web sites for modifying or updating the content or structure.

6.5.1   Adaptive Web Sites

Adaptive Web Site are systems which adapt their content, structure and/or presentation of the Web Objects, to each individual user's characteristics, usage behaviour and/or usage environment [39].Adaptive sites provide users with both personalized and recommended services, and content according to the user's profile acquired by the system [28, 83]. The server load can be optimized since hyperlink demands can be forecasted and automatic balances could be performed. The topology of the web site can be modified to the web user interest. Different aspects of managing a web site are benefited from this technology. Marketing purposes have highly improved for adaptive sites. Usability trends are solved by means of specific user requirement.

6.5.2   Web Personalization

Web Personalization improves the web site structure based on the interaction with all visitors. Profiling is the principal processing that must be performed for those purposes. Using the profiling information, an automatic classification of web user should return the profile association with an objective (e.g. product). There are two kind of personalization [51, 83] based on the degree of conflict with the current semantic of the web site.

Tactical adaptation: It does not affect the overall structure of the web site as a semantic consistency is maintained and can be implemented by automatic systems. Such kinds of systems are autonomous and the whole design of the web site contemplates dynamic changes. Web sites like Amazon, NetFlix and others implement this kind of personalization.

Strategic adaptation: It must have the agreement of the owner of the web site, since the suggested changes are in conflict with the original orientation of the web site. Off-line recommendations are in general used for this kind of adaptation where owner's feedback is part of the component of the process [63].

6.5.3   Recommendation

Recommendation is also based on profiling and its objective is to retrieve a product that is most likely to be selected by the current web user. Furthermore, the profiling processing for web user's usage can be described from two different points of view depending on whether the profile is pre-established or is created on the run. The general process is called "Filtering", which is described depending on the orientation [40].

Content Filtering: Categories are created according to the nature of the Web User. The information about the user is retrieved from customer databases and associations with objectives like promotions or products are performed by a trained classifier.

Collaborative Filtering: The profile is created on the run based on the history of user's interaction with the web site. The technique relates to the associations between products and relationship with users history. Some approaches relate to the product's neighbourhood in which similar rating on other products is provided by other similar users. This technique is called Neighbourhood Method, where categories clusters are defined by the user past product rating. For instance a book entitled "Calculus" is visited in the neighbour of Authors like "Spivak" or "Apostol" as web users visit a web book selling site. Latent Factor Model is another approach based on recognizing factor variables that help to measure how useful a product is for a user. Once the factor is discovered, the importance of the produce is determined for a user. Matrix factorization methods have been used for implementing Latent Factor Models like in the 2009 winner NetFlix price [46].

6.6 Summary

Web Usage Mining has been studied for more than ten year with successful application. Nevertheless, vast quantity of new research in the area of applied behavioural sciences and other new trends such as matrix factorization methods have revolutionised the field of web mining. NetFlix price has demonstrated that traditional web usage mining techniques have little impact on real world issues. However, recent advances in this field have reaped promising results.

Acknowledgements. This work was supported partially by the FONDEF project DO8I-1015, the *National Doctoral Grant* from *Conicyt Chile* and the *Web Intelligence Research Group* (wi.dii.uchile.cl) is greatly acknowledged.

References

1. Ayyagari, P., Sun, Y.: Modeling the internet and the web: Probabilistic methods and algorithms. by pierre baldi, paolo frasconi, padhraic smith, john wiley and sons ltd., west sussex, england, 2003. 285 pp isbn 0 470 84906 1. Inf. Process. Manage. 42(1), 325–326 (2006)
2. Baeza-yates, R., Castillo, C.: Crawling the infinite web: Five levels are enough. In: Leonardi, S. (ed.) WAW 2004. LNCS, vol. 3243, pp. 156–167. Springer, Heidelberg (2004)
3. Berendt, B., Hotho, A., Stumme, G.: Data preparation for mining world wide web browsing patterns. Journal of Knowlegde and Information Systems 1(1), 5–32 (1999)

4. Berendt, B., Mobasher, B., Spiliopoulou, M., Wiltshire, J.: Measuring the accuracy of sessionizers for web usage analysis. In: Proc. of the Workshop on Web Mining, First SIAM Internat.Conf. on Data Mining, pp. 7–14 (2001)
5. Bixby, R.E.: Solving real-world linear programs: A decade and more of progress. Operations Research 50(1), 3–15 (2002)
6. Borges, J., Levene, M.: Data mining of user navigation patterns. In: WEBKDD 1999: Revised Papers from the International Workshop on Web Usage Analysis and User Profiling, London, UK, pp. 92–111. Springer, Heidelberg (2000)
7. Borges, J., Levene, M.: Evaluating variable-length markov chain models for analysis of user web navigation sessions. IEEE Trans. on Knowl. and Data Eng. 19(4), 441–452 (2007)
8. Burget, R., Rudolfova, I.: Web page element classification based on visual features. In: Asian Conference on Intelligent Information and Database Systems, vol. 0, pp. 67–72 (2009)
9. Cadez, I., Heckerman, D., Meek, C., Smyth, P., White, S.: Visualization of navigation patterns on a web site using model-based clustering. In: KDD 2000: Proceedings of the sixth ACM SIGKDD international conference on Knowledge discovery and data mining, pp. 280–284. ACM, New York (2000)
10. Chakrabarti, D., Faloutsos, C.: Graph mining: Laws, generators, and algorithms. ACM Comput. Surv. 38(1), 2 (2006)
11. Chen, X., Zhang, X.: A popularity-based prediction model for web prefetching. Computer 36(3), 63–70 (2003)
12. Cooley, R., Mobasher, B., Srivastava, J.: Towards semantic web mining. In: Proc. in First Int. Semantic Web Conference, pp. 264–278 (2002)
13. Dell, R.F., Román, P.E., Velásquez, J.D.: Web user session reconstruction using integer programming. In: Procs. of The 2008 IEEE/WIC/ACM International Conference on Web Intelligence, Sydney, Australia, December 2008, pp. 385–388 (2008)
14. Demir, G.N., Uyar, A.S., Ögüdücü, S.G.: Graph-based sequence clustering through multiobjective evolutionary algorithms for web recommender systems. In: GECCO 2007: Proceedings of the 9th annual conference on Genetic and evolutionary computation, pp. 1943–1950. ACM, New York (2007)
15. Dongshan, X., Junyi, S.: A new markov model for web access prediction. Computing in Science and Engg. 4(6), 34–39 (2002)
16. Dujovne, L.E., Velásquez, J.D.: Design and implementation of a methodology for identifying website keyobjects. In: Velásquez, J.D., Ríos, S.A., Howlett, R.J., Jain, L.C. (eds.) Knowledge-Based and Intelligent Information and Engineering Systems. LNCS, Part I, vol. 5711, pp. 301–308. Springer, Heidelberg (2009)
17. Eirinaki, M., Vazirgiannis, M., Kapogiannis, D.: Web path recommendations based on page ranking and markov models. In: WIDM 2005: Proceedings of the 7th annual ACM international workshop on Web information and data management, pp. 2–9. ACM, New York (2005)
18. Facca, F., Lanzi, P.: Recent developments in web usage mining research. In: Kambayashi, Y., Mohania, M., Wöß, W. (eds.) DaWaK 2003. LNCS, vol. 2737, pp. 140–150. Springer, Heidelberg (2003)
19. Felzenszwalb, P.F., Huttenlocher, D.P., Kleinberg, J.M.: Fast algorithms for large-state-space hmms with applications to web usage analysis. In: Thrun, S., Saul, L.K., Schölkopf, B. (eds.) NIPS. MIT Press, Cambridge (2003)
20. Glassman, S.: A caching relay for the world wide web. In: Selected papers of the first conference on World-Wide Web, pp. 165–173. Elsevier Science Publishers B. V., Amsterdam (1994)

21. Grannis, K., Davis, E.: Online sales to climb despite struggling economy, According to Shop.org/Forrester Research Study (2008)
22. Grannis, K., Davis, E.: China internet network information center. In: 14th statistical survey report on the internet development of china (2009),
http://www.cnnic.net.cn/uploadfiles/pdf/2009/10/13/94556.pdf
23. Gündüz, Ş., Özsu, M.T.: A web page prediction model based on click-stream tree representation of user behavior. In: KDD 2003: Proceedings of the ninth ACM SIGKDD international conference on Knowledge discovery and data mining, pp. 535–540. ACM, New York (2003)
24. Guo, Y.Z., Ramamohanarao, K., Park, L.A.: Grouped ecoc conditional random fields for prediction of web user behavior. In: Theeramunkong, T., Kijsirikul, B., Cercone, N., Ho, T.-B. (eds.) PAKDD 2009. LNCS, vol. 5476, pp. 757–763. Springer, Heidelberg (2009)
25. Guo, Y.Z., Ramamohanarao, K., Park, L.A.F.: Web page prediction based on conditional random fields. In: Proceeding of the 2008 conference on ECAI 2008, pp. 251–255. IOS Press, Amsterdam (2008)
26. Hasan, T., Mudur, S.P., Shiri, N.: A session generalization technique for improved web usage mining. In: WIDM 2009: Proceeding of the eleventh international workshop on Web information and data management, pp. 23–30. ACM, New York (2009)
27. Hay, B., Wets, G., Vanhoof, K.: Mining navigation patterns using a sequence alignment method. Knowl. Inf. Syst. 6(2), 150–163 (2004)
28. Hongwei, W., Xie, L.: Adaptive site design based on web mining and topology. In: CSIE 2009: Proceedings of the 2009 WRI World Congress on Computer Science and Information Engineering, Washington, DC, USA, pp. 184–189. IEEE Computer Society, Los Alamitos (2009)
29. Huberman, B., Pirolli, P., Pitkow, J., Lukose, R.M.: Strong regularities in world wide web surfing. Science 280(5360), 95–97 (1998)
30. Huntington, P., Nicholas, D., Jamali, H.R.: Website usage metrics: A re-assessment of session data. Information Processing & Management 44(1), 358–372 (2008)
31. Hannah Inbarani, H., Thangavel, K., Pethalakshmi, A.: Rough set based feature selection for web usage mining. In: ICCIMA 2007: Proceedings of the International Conference on Computational Intelligence and Multimedia Applications (ICCIMA 2007), Washington, DC, USA, pp. 33–38. IEEE Computer Society, Los Alamitos (2007)
32. Ingwersen, P., Jirvelin, K.: The Turn: Integration of Information Seeking and Retrieval in Context, 1st edn. Springer, Heidelberg (2005)
33. Jespersen, S., Pedersen, T.B., Thorhauge, J.: Evaluating the markov assumption for web usage mining. In: WIDM 2003: Proceedings of the 5th ACM international workshop on Web information and data management, pp. 82–89. ACM, New York (2003)
34. Jin, X., Zhou, Y., Mobasher, B.: Web usage mining based on probabilistic latent semantic analysis. In: KDD 2004: Proceedings of the tenth ACM SIGKDD international conference on Knowledge discovery and data mining. ACM, New York (2004)
35. Joshi, A., Krishnapuram, R.: On mining web access logs. In: Proc. of the 2000 ACM SIGMOD Workshop on Research Issue in Data Mining and Knowledge Discovery, pp. 63–69 (2000)
36. Juels, A., Jakobsson, M., Jagatic, T.N.: Cache cookies for browser authentication (extended abstract). In: SP 2006: Proceedings of the 2006 IEEE Symposium on Security and Privacy, Washington, DC, USA, pp. 301–305. IEEE Computer Society, Los Alamitos (2006)
37. Kausshik, A.: Web Analytics 2.0: The Art of Online Accountability and Science of Customer Centricity. Sybex (2009)

38. Kellar, M.: An Examination of User Behaviour during Web Information Tasks. PhD thesis, Dalhousie University, Halifax, Nova Scotia, Canada (2007)
39. Kolias, C., Kolias, V., Anagnostopoulos, I., Kambourakis, G., Kayafas, E.: Enhancing user privacy in adaptive web sites with client-side user profiles. In: SMAP 2008: Proceedings of the 2008 Third International Workshop on Semantic Media Adaptation and Personalization, Washington, DC, USA, pp. 170–176. IEEE Computer Society, Los Alamitos (2008)
40. Koren, Y., Bell, R., Volinsky, C.: Matrix factorization techniques for recommender systems. Computer 42(8), 30–37 (2009)
41. Lafferty, J.D., McCallum, A., Pereira, F.C.N.: Conditional random fields: Probabilistic models for segmenting and labeling sequence data. In: ICML 2001: Proceedings of the Eighteenth International Conference on Machine Learning, pp. 282–289. Morgan Kaufmann Publishers Inc., San Francisco (2001)
42. Lan, M., Tan, C.L., Su, J., Lu, Y.: Supervised and traditional term weighting methods for automatic text categorization. IEEE Trans. Pattern Anal. Mach. Intell. 31(4), 721–735 (2009)
43. Langford, D.: Internet ethics. MacMillan Press Ltd., Basingstoke (2000)
44. Levene, M., Borges, J., Loizou, G.: Zipf's law for web surfers. Knowl. Inf. Syst. 3(1), 120–129 (2001)
45. Liu, J., Zhang, S., Yang, J.: Characterizing web usage regularities with information foraging agents. IEEE Trans. on Knowl. and Data Eng. 16(5), 566–584 (2004)
46. Lohr, S.: A 1 million dollars research bargain for netflix, and maybe a model for others. New York Times (2009)
47. Loyola, P., Román, P.E., Velásquez, J.D.: Colony surfer: Discovering the distribution of text preferences from web usage. In: Procs. Of the First Workshop in Business Analytics and Optimization, BAO (2010)
48. Manning, C.D., Schutze, H.: Fundation of Statistical Natural Language Processing. The MIT Press, Cambridge (1999)
49. Masseglia, F., Poncelet, P., Teisseire, M., Marascu, A.: Web usage mining: extracting unexpected periods from web logs. Data Min. Knowl. Discov. 16(1), 39–65 (2008)
50. Mayer-Schonberger, V.: Nutzliches vergessen. In: Goodbye privacy grundrechte in der digitalen welt (Ars Electronica), pp. 253–265 (2008)
51. Mikroyannidis, A., Theodoulidis, B.: Heraclitus: A framework for semantic web adaptation. IEEE Internet Computing 11(3), 45–52 (2007)
52. Mobasher, B., Cooley, R., Srivastava, J.: Creating adaptive web sites through usage-based clustering of urls. In: KDEX 1999: Proceedings of the 1999 Workshop on Knowledge and Data Engineering Exchange, Washington, DC, USA, p. 19. IEEE Computer Society, Los Alamitos (1999)
53. Mobasher, B., Cooley, R., Srivastava, J.: Automatic personalization based on web usage mining. Commun. ACM 43(8), 142–151 (2000)
54. Mobasher, B., Dai, H., Luo, T., Nakagawa, M.: Effective personalization based on association rule discovery from web usage data. In: WIDM 2001: Proceedings of the 3rd international workshop on Web information and data management, pp. 9–15. ACM, New York (2001)
55. Mobasher, B., Dai, H., Luo, T., Nakagawa, M.: Discovery and evaluation of aggregate usage profiles for web personalization. Data Min. Knowl. Discov. 6(1), 61–82 (2002)
56. Obendorf, H., Weinreich, H., Herder, E., Mayer, M.: Web page revisitation revisited: implications of a long-term click-stream study of browser usage. In: CHI 2007: Proceedings of the SIGCHI conference on Human factors in computing systems, pp. 597–606 (2007)

57. Olston, C., Chi, E.H.: Scenttrails: Integrating browsing and searching on the web. ACM Trans. Comput.-Hum. Interact. 10(3), 177–197 (2003)
58. Park, S., Suresh, N.C., Jeong, B.-K.: Sequence-based clustering for web usage mining: A new experimental framework and ann-enhanced k-means algorithm. Data Knowl. Eng. 65(3), 512–543 (2008)
59. Paterek, A.: Improving regularized singular value decomposition for collaborative filtering. In: Proceedings of KDD Cup and Workshop (2007)
60. Perkowitz, M., Etzioni, O.: Towards adaptive web sites: conceptual framework and case study. Artif. Intell. 118(1-2), 245–275 (2000)
61. Plumbaum, T., Stelter, T., Korth, A.: Semantic web usage mining: Using semantics to understand user intentions. In: Houben, G.-J., McCalla, G., Pianesi, F., Zancanaro, M. (eds.) UMAP 2009. LNCS, vol. 5535, pp. 391–396. Springer, Heidelberg (2009)
62. Resnick, S.I.: Adventures in stochastic processes. Birkhauser Verlag, Basel (1992)
63. Ríos, S.A., Velásquez, J.D.: Semantic web usage mining by a concept-based approach for off-line web site enhancements. In: WI-IAT 2008: Proceedings of the 2008 IEEE/WIC/ACM International Conference on Web Intelligence and Intelligent Agent Technology, Washington, DC, USA, pp. 234–241. IEEE Computer Society, Los Alamitos (2008)
64. Román, P.E., Velásquez, J.D.: Markov chain for modeling web user browsing behavior: statistical inference. In: XIV Latin Ibero-American Congress on Operations Research, CLAIO (2008)
65. Román, P.E., Velásquez, J.D.: Analysis of the web user behavior with a psychologically-based diffusion model. In: Procs. of the AAAI 2009 Fall Symposium on Biologically Inspired Cognitive Architectures, Arlington VA, USA, p. 72 (2009)
66. Román, P.E., Velásquez, J.D.: A dynamic stochastic model applied to the analysis of the web user behavior. In: The 2009 AWIC 6th Atlantic Web Intelligence Conference, Prague, Czech Republic, pp. 31–40 (2009) (Invited Lecture)
67. Runkler, T.A., Bezdek, J.C.: Web mining with relational clustering. International Journal of Approximate Reasoning 32(2-3), 217–236 (2003)
68. Brin, S., Page, L.: The anatomy of a large-scale hypertextual web search engine. In: Computer Networks and ISDN Systems, pp. 107–117 (1998)
69. Sarukkai, R.R.: Link prediction and path analysis using markov chains. In: Proceedings of the 9th international World Wide Web conference on Computer networks: the international journal of computer and telecommunications networking, pp. 377–386. North-Holland Publishing Co., Amsterdam (2000)
70. Schmidt, K.-U., Stojanovic, L., Stojanovic, N., Thomas, S.: On enriching ajax with semantics: The web personalization use case. In: Franconi, E., Kifer, M., May, W. (eds.) ESWC 2007. LNCS, vol. 4519, pp. 686–700. Springer, Heidelberg (2007)
71. Sen, R., Hansen, M.: Predicting web user's next access based on log data. J. Comput. Graph. Stat. 12(1), 143–155 (2003)
72. Shi, J., Shi, F., Qiu, H.: User's interests navigation model based on hidden markov model. In: Wang, G., Liu, Q., Yao, Y., Skowron, A. (eds.) RSFDGrC 2003. LNCS (LNAI), vol. 2639, pp. 631–634. Springer, Heidelberg (2003)
73. Snásel, V., Kudelka, M.: Web content mining focused on named objects. In (IHCI) First International Conference on Intelligent Human Computer Interaction, pp. 37–58. Springer, India (2009)
74. Spiliopoulou, M., Mobasher, B., Berendt, B., Nakagawa, M.: A framework for the evaluation of session reconstruction heuristics in web-usage analysis. Informs Journal on Computing 15(2), 171–190 (2003)

75. Spiliopoulou, M., Faulstich, L.: Wum: A web utilization miner. In: Atzeni, P., Mendelzon, A.O., Mecca, G. (eds.) WebDB 1998. LNCS, vol. 1590, pp. 184–203. Springer, Heidelberg (1998)
76. Srivastava, J., Cooley, R., Deshpande, M., Tan, P.: Web usage mining: Discovery and applications of usage patterns from web data. SIGKDD Explorations 2(1), 12–23 (2000)
77. Takács, G., Piliászy, I., Németh, B., Tikk, D.: Major components of the gravity recommendation system. SIGKDD Explor. Newsl. 9(2), 80–83 (2007)
78. Tao, Y.-H., Hong, T.-P., Lin, W.-Y., Chiu, W.-Y.: A practical extension of web usage mining with intentional browsing data toward usage. Expert Syst. Appl. 36(2), 3937–3945 (2009)
79. Tao, Y.-H., Hong, T.-P., Su, Y.-M.: Web usage mining with intentional browsing data. Expert Syst. Appl. 34(3), 1893–1904 (2008)
80. Vazquez, A., Gama Oliveira, J., Dezso, Z., Goh, K.-I., Kondor, I., Barabasi, A.-L.: Modeling bursts and heavy tails in human dynamics. Physical Review E 73(3), 36127 (2006)
81. Vazquez, A.: Exact results for the barabasi model of human dynamics. Physical Review Letters 95(24), 248701 (2005)
82. Velásquez, J.D., Yasuda, H., Aoki, T., Weber, R.: A new similarity measure to understand visitor behavior in a web site. IEICE Transactions on Information and Systems, Special Issues in Information Processing Technology for web utilization E87-D(2), 389–396 (2004)
83. Velásquez, J.D., Palade, V.: Adaptive web sites: A knowledge extraction from web data approach. IOS Press, Amsterdam (2008)
84. Velásquez, J.D., Estévez, P.A., Yasuda, H., Aoki, T., Vera, E.S.: Intelligent web site: Understanding the visitor behavior. In: Negoita, M.G., Howlett, R.J., Jain, L.C. (eds.) KES 2004. LNCS (LNAI), vol. 3213, pp. 140–147. Springer, Heidelberg (2004)
85. Velásquez, J.D., Weber, R., Yasuda, H., Aoki, T.: A methodology to find web site keywords. In: EEE 2004: Proceedings of the 2004 IEEE International Conference on e-Technology, e-Commerce and e-Service (EEE 2004), Washington, DC, USA, pp. 285–292. IEEE Computer Society, Los Alamitos (2004)
86. Wang, S., Gao, W., Huang, T., Ma, J., Li, J., Xie, H.: Adaptive online retail web site based on hidden markov model. In: Lu, H., Zhou, A. (eds.) WAIM 2000. LNCS, vol. 1846, pp. 177–188. Springer, Heidelberg (2000)
87. White, R.W.: Investigating behavioral variability in web search. In: Proc. WWW, pp. 21–30 (2007)
88. White, R.W., Drucker, S.M.: Investigating behavioral variability in web search. In: WWW 2007: Proceedings of the 16th international conference on World Wide Web (2007)
89. Wolpert, D.H.: Stacked generalization. Neural Networks 5, 241–259 (1992)
90. Xexeo, G., de Souza, J., Castro, P.F., Pinheiro, W.A.: Using wavelets to classify documents. In: IEEE/WIC/ACM International Conference on Web Intelligence and Intelligent Agent Technology, vol. 1, pp. 272–278 (2008)
91. Xiao, J., Zhang, Y., Jia, X., Li, T.: Measuring similarity of interests for clustering web-users. In: ADC 2001: Proceedings of the 12th Australasian database conference, Washington, DC, USA, pp. 107–114. IEEE Computer Society, Los Alamitos (2001)
92. Zhou, Y., Leung, H., Winoto, P.: Mnav: A markov model-based web site navigability measure. IEEE Trans. Softw. Eng. 33(12), 869–890 (2007)
93. Zukerman, I., Albrecht, D.W., Nicholson, A.E.: Predicting users' requests on the www. In: UM 1999: Proceedings of the seventh international conference on User modeling, Secaucus, NJ, USA, pp. 275–284. Springer-Verlag New York, Inc. (1999)

Chapter 7
User-Centric Web Services for Ubiquitous Computing

In-Young Ko, Hyung-Min Koo, and Angel Jimenez-Molina

Abstract. One of the key issues in ubiquitous computing environments is to realize user centricity, which is about enabling Web-based services and services in general to be prepared, supported and delivered from the perspective of users rather than system elements. To accomplish user centricity, it is essential to deal with the high-level goals of multiple users and with dynamic changes of contextual information. The approach of task-oriented computing is a means of realizing a user-centric service provision in Ubiquitous Computing environments. This approach represents users' goals in tasks, which are then bound to available service instances. This chapter describes the essential requirements for a user-centric service provision in Ubiquitous Computing environments. User centricity, context awareness, composability and dynamicity are such requirements. Existing works regarding service provision in Ubiquitous Computing are analyzed by highlighting their limitations in the light of those requirements. This chapter also describes a service framework for Ubiquitous Computing. This framework consists of three layers - the task layer, the composition-pattern layer and the service layer. The essential semantic elements that describe a task are arranged in a task description model. That model is used by a semantically based mechanism to select appropriate tasks based on context information. It is also utilized by a pattern-based service composition mechanism. The framework is illustrated by a demo application example developed in our test bed. By leveraging the task-oriented computing approach, this chapter contributes to enhancing the reusability of tasks and patterns, improving flexibility when developing Ubiquitous Computing applications and providing dynamism to the service provision in this type of environments.

In-Young Ko · Hyung-Min Koo · Angel Jimenez-Molina
KAIST, Korea Advanced Institute of Science and Technology,
335 Gwahangno (373-1 Guseong-dong), Yuseong-gu,
Daejeon (305-701), Republic of Korea
e-mail: iko@kaist.ac.kr, hmkoo@kaist.ac.kr, anjimenez@kaist.ac.kr

J.D. Velásquez and L.C. Jain (Eds.): Advanced Techniques in Web Intelligence – 1, SCI 311, pp. 167–189.
springerlink.com   © Springer-Verlag Berlin Heidelberg 2010

7.1   Introduction

Ubiquitous Computing (Ubicomp) is a paradigm shift that enables users to access
networked and Web-based services anywhere and anytime [22, 34]. It leads to invisible Web-based services that transparently support users daily life routines.

Providing Web-based services in Ubicomp environments leads to a set of distinguishable requirements - *user centricity* , *context awareness* [18], *composability of services* , and *dynamicity* of user-goals and environmental conditions. *User
centricity* enables services to be prepared, supported and delivered from the perspective of user goals rather than system elements. User centricity is very important for either individual user or multi-user satisfaction in Ubicomp environments.
Context-Awareness reflects the context information in selecting appropriate Web-
based services and services in general. The context information characterizes the
state of the users and their surrounding Ubicomp environments. Examples of context information include locations and physiological states - such as the heartbeat or
blood pressure - of users, as well as the temperature, brightness and humidity of the
surrounding environment. *Composability* is a requirement that makes it easier to develop Ubicomp applications by coordinating a set of Ubicomp services during runtime operation. Such a composition must be carried out according to a user-centric
viewpoint and contextual information. *Dynamicity* considers changes of user goals
and context information when realizing and executing Web-based services and services in general. These dynamic factors need to be reflected in supporting users that
move among multiple Ubicomp spaces by tailoring useful combinations of services
and by dynamically reconfiguring the combinations of services according to those
changes.

In Ubicomp environments, it is critical to map the high-level goals that are represented in users' perspectives onto Web-based services that are defined from the perspective of the system environment. In addition, user goals should be dynamically
supported through the utilization of services that are available in a local environment
and on the Web.

The major areas of research in providing services in Ubicomp environments
include the ABC [6], the IST Amigo [17], the Gaia [21], and the Aura [24],
projects. These studies have been successful in discovering, configuring, and providing the services that are needed to support users goals. However, further research
on this area should focus on bridging the gap between users' goals and services
dynamically.

The concept of task-oriented computing [33] realizes user centricity, context
awareness, composability, and dynamicity by representing users' goals in tasks ,
which are then mapped to Web-based services and available services in a Ubicomp
environment based on context information. Task-oriented computing bridges the gap
between tasks and services, taking into account their different perspectives, granularities and abstraction levels [9, 10, 11, 12].

In order to realize the service-oriented paradigm in Ubicomp properly and
overcome the limitations, a task-oriented service framework for Ubicomp is described [9, 10]. This framework is composed of three layers - *a task layer*, a

composition-pattern layer, and a *service layer*. The *task layer* includes definitions of tasks that support users' goals. Each task definition is composed of a set of coordinated actions . In the *composition-pattern layer*, a set of required services (termed abstract services) and the associated coordination logic are represented in a service composition-pattern (SCP) . A SCP maps an action, which is defined from the perspective of users, to the Web-based services and available service instances that are tightly bound to the local environment. The *service layer* is composed of Web-based services, and services that are deployed in an environment. Services are normally bound with local devices and Web-based computational resources and deliver predefined functionality. The framework and its embedded mechanisms were effective for realizing and reconfiguring tasks during demonstrations when implemented in our test bed and in experimental trials conducted with several combinations of context information taken from actual datasets of user activities.

A roadmap of this chapter is as follows. Section 7.2 provides the essential requirements for providing Web-based services in Ubicomp environments. Current research in Ubicomp services is described in Section 7.3. This related work is analyzed in terms of the essential requirements for service provisioning in Ubicomp. Section 7.4 explains a task-oriented service framework for Ubicomp that fully reflects those requirements. Section 7.5 concludes the chapter.

7.2 Essential Requirements for Providing Web Services in Ubicomp

The major requirements for service provisioning in Ubicomp environments correspond to user centricity, context awareness, composability and dynamicity. User centricity in turn requires transparency and multiple-user support , while dynamicity requires multi-space support and reconfigurability.

7.2.1 User Centricity

Services need to be prepared and delivered from the perspective of users rather than systems. This separation of concern between users and systems is termed user centricity. This requirement generates the goal of capturing, in a certain way, knowledge about user needs. This knowledge must be used to recognize the user intentions under diverse contextual situations.

The approach of task-oriented computing supports the goal of capturing users' needs. This approach realizes the perspective of users by representing their high-level goals in tasks [33]. Therefore, it is necessary to reflect the essential characteristics of users and Ubicomp environments in the description of tasks. That description needs to be used to select appropriate tasks according to the contextual information. Additionally, in order to deliver user-centric services a process to bridge the gap between the user-centric task view and services available in the Ubicomp environment

or the Web is required. Two major requirements for realizing user centricity in a
Ubicomp environment are transparency and multi-user support.

7.2.1.1   Transparency

Regarding transparency, an essential requirement of Ubicomp is to free users from
their concerns about the existence of specific services in an environment and
whether certain services can be accessed on the Web. Additionally, transparency
involves simplifying the process of selecting for users, and composing then delivering those services [8]. It ensures that any technical detail related to system concerns
that could prevent novice users from completing their tasks in a Ubicomp environment are transparently managed by the underlying infrastructure. That is, tasks are
transparently selected and mapped to available services without user intervention.

7.2.1.2   Multi-user Suppport

User centricity is not only about satisfying the needs of a single user, but also taking
into account the needs of users that share a presence in a Ubicomp environment.
Thus, it is necessary to recognize the goal of a group of users engaged in a collective task, and support them with services accordingly. Unlike collaborative goals,
multiple groups or individual users could be participating in different tasks with
conflicting goals. For instance, a group of people working together on a research
project might require the room to be quiet from external noise, thus a window closing service would need to be activated. However, a user entering the same room at
the same moment after an exercise session would prefer to open a window in order
to make the room cooler. The basic and traditional means of addressing this problem of opposing goals is to resolve the conflict as it is detected when the services
instances to be executed are selected. Nevertheless, a more effective way would be
to predict the conflict in advance at the task level, thus making the room cooler by
turning on the air conditioner and closing the window.

7.2.2   Context Awareness

A selection of appropriate tasks needs to reflect the contextual information that
characterizes the state of a user or a group of users and the Ubicomp environment.
In general, reflecting context information to instantiate appropriate applications is
known as context awareness. This is an active research issue in Ubicomp [1, 2, 4, 5].
Examples of the context information include *locations* (e.g., a living room, bed-
room, kitchen, gym, or waiting room), *profiles* (e.g., interest, gender, age group,
religion, marital status, educational information, or occupation), or the *physiological states* (e.g., heartbeat or blood pressure) of users; the *temperature, brightness,
humidity* or *social constraints* of the surroundings environment, and the *day of the
week* or *hour of the day*, among others contexts.

The relevance of context awareness for a user-centric service provision is that
the state of a user engaged in a task and its surrounding environment, among other

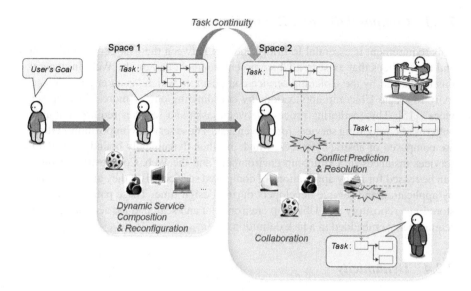

Fig. 7.1 Major Requirements of a User-centric Web Service Provision in Ubicomp

context values, must be considered in the service selection process, specifically in guiding the sequence of selection factors and in weighting up their importance [11]. According to recent works [19] on the subject of task recognition in Ubicomp environments, the location and time are among the best factors for predicting the most likely activities to be performed by a user. Thus, a task-selection mechanism should prioritize these factors. As for temporal context, certain tasks are clearly more frequent on specific days of the week and at certain hours of the day [19]. On the other hand, certain tasks are most likely to be performed in specific *locations*: a research discussion meeting would most likely be held in a conference room on a campus. Other factors should also be considered, but mainly to make it more specialized the general selection of tasks that the *location* and *time* properties would produce. For instance, having a set of task candidates determined by *location* and *time properties*, the similarity of the *age range* among a group of teenagers would specify a set of more specialized tasks. An example of these more specialized tasks is socializing such as *dancing*.

Additionally, *social and spatial constraints* attached to a *location* would restrict the possible tasks that can be performed there, or the manner in which a task can be realized with specific services. Examples of such constraints are the permitted *level of noise* in the campus corridors; *brightness* in the auditorium, or *social values* that must be guaranteed in a bathroom, such as *privacy* and *individuality*. According to those constraints, a *play music service* that uses a public speaker or a *video record service* should not be activated in a campus corridor and a bathroom, respectively.

7.2.3   Composability and Reusability

This requirement is essential to bridge the gap between the user-centric task view and the services that exist in a Ubicomp environment and on the Web. In general, this is realized by the Ubicomp service composition, which "is an engineering activity to create Ubicomp applications by coordinating one or more Ubicomp services" [15]. This mediating process between tasks and services requires common patterns of configuring services for a given task. Those service composition-patterns are composed of abstract services, which are then dynamically bound to available services instances in a Ubicomp environment and the Web. Composition-patterns can be reused for many applications of composed services [10]; they are determined by application and service developers during design time. These patterns must be stored in a repository in a Ubicomp environment and searched for according to the requirements embedded in a task definition.

7.2.4   Dynamicity

7.2.4.1   Multi-space Suppport

Users move across multiple Ubicomp environments. Thus, it is necessary to support user tasks transparently across those environments. It is essential to ensure task continuity without any delay or user intervention. To accomplish the task continuity goal, the status information pertaining to a task needs to be exchanged among Ubicomp environments. For instance, for a user watching a movie at his home, if the user moves to another Ubicomp environment, the same movie could be continued. That is, the user could receive the options of continuing the movie at the position where it was stopped, or to rewind the movie in order to remember the last part he/she watched. In order to realize this type of task continuity, it is necessary to verify the availability of appropriate resources in the new environment in which the functionality needs to be equivalent to that in the initial space.

7.2.4.2   Reconfigurability

A major requirement of Ubicomp is to improve application adaptability to the dynamic changes that can occur in a Ubicomp environment. That is, changes in context information and user goals require the reconfiguration of tasks during runtime. Since a task is configured by a set of coordinated SCPs, such a reconfiguration can be carried out by adding, substituting, or deleting SCPs over a task. Therefore, a SCP is a unit that is used for the dynamic reconfiguration of a Ubicomp application during runtime. Reconfiguring SCPs is more effective for reflecting changes than directly reconfiguring current services. In this way, SCPs can be composed dynamically to meet changes on user requirements and contextual information. This needs to be done by selecting the SCPs where functionality is semantically equivalent to the functionality that realizes the new user requirements and context.

Regarding the requirements of reconfiguration in task continuity, they may involve a reconfiguration of a task that is being resumed according to the new contextual information and available services in the environment. Therefore, if the user moves from his home to a waiting room at a hospital, the task of watching a movie instantiates a search for the movie file on the Web and activates the public screen of the environment to show it if no other users are in the waiting room. The task must be smoothly switched to the users mobile device if a new user enters the room so as not to disturb that person. In such a case, SCPs for deploying media content should exist either for public devices or personal devices.

7.3   Current Research in Ubicomp Web Services

Many researchers have tried to identify user activities in an accurate and efficient manner by taking into account the contextual information of a Ubicomp environment [11]. Moreover, the requirements described in this chapter complement the future research roadmap that should be carried out on service provailing in Ubicomp environments. Therefore, the future research should be addressed by the following issues: (1) separation of concerns between a user-centric task view and a system perspective, reflecting them at different abstraction levels; (2) selection of coarse-grained tasks based on dynamic context information; (3) dynamic service composition and reconfiguration at an abstract level instead of managing current services. Hence, further investigation is needed to address the requirements for providing Web-based services and services in general in Ubicomp.

7.3.1   Gaia

Gaia [21] is a user-centric, location-aware and event-driven framework for building Ubicomp applications. Gaia provides a mechanism for building new applications and for making use of a predefined application model. This model consists of an ontology of generic applications that is used to generate customized applications in an Ubicomp environment - termed an *active space* in Gaia's nomenclature.

Each generic application is described by means of an *Application General Description* (AGD). In order to instantiate a new application, the framework extends an existing AGD by creating an *Application Customized Description* (ACD). Application instantiation is done by gathering the user information in an active space. This information consists of the user requirements, goals, interests, preferences, and other personalized factors. Gaia then searches for an AGD that properly reflects the user information. This AGD is customized in an ACD. Components of such a customized application can be reconfigured during runtime according to changes in environments. Additionally, users must make their information available, even when they move across different active spaces. If it is necessary to continue the execution of a given application when a user switches to a new active space, semantically similar applications are selected. That similarity is measured in terms of functionality and behavior to ensure a seamless experience to the user in relation to the service provision. The ACD from the original active space is similar to a "snapshot" of the application, which also embeds its state. In the new active space, a new ACD model is created. This feature is termed a *mobile polymorphic application* [20] in the Gaia framework. It allows applications to change their structure dynamically given inter-space user migration.

One of the limitations of this framework involves the static binding of the user information within the predefined and available applications in a Ubicomp environment. In addition, Gaia uses an approach based on a system perspective coupled with ACD descriptions rather than a user perspective.

7.3.2 Aura

Aura [24] adopts a task-oriented computing approach for a user-centric service provision in Ubicomp environments. In Aura, user centricity is realized by describing user goals in coarse-grained tasks. As stated by Wang and Garlan [33], "users can interact with the system in terms of high-level tasks instead of individual services or applications." Tasks are dynamically mapped to virtual services, which are in turn associated with actual devices or applications while maximizing user utility.

Although Aura is a representative of a task-oriented framework, it assumes a given knowledge of the relationships between virtual services and concrete service suppliers. This limitation implies a static binding relationship between different levels of abstraction and granularities.

7.3.3 ABC Framework

The ABC framework [6] focuses on adapting well-established and stable tasks to heterogeneous resources in Ubicomp environments. These resources are composed of a set of predefined, task-centered applications, services and other computational elements. Two major issues highlighted in this research are collaboration among users and context awareness. The former issue deals with identifying and managing the activities that are shared amongst several participants, while the latter is concerned with enabling activities to adapt to changes in contextual information.

The ABC framework does not properly realize some of the requirements for user-centric service provisioning in the Ubicomp environments discussed in this chapter. First, although it adopts a task-oriented framework, it mostly deals with activities that are related to small computations: tasks are defined in a fine-grained approach. Fine-grained tasks do not enable developers to make Ubicomp applications based on high-level user goals, which must be represented in coarse-grained tasks. Second, this framework does not declare its capability to support inter-space seamless task execution.

7.3.4   IST Amigo

The IST Amigo framework [17] is an example of a task-oriented Ubicomp service provision in ambient systems domain. Ambient systems are of interest as they pertain to the present study, as they share a number of identical features with Ubicomp environments. Both attempt to support user applications transparently with user goals in an environment, and both attempt to meet the requirement of user-centricity. This framework makes use of predefined descriptions of applications in the shape of workflows, which represent user goals. Workflows can be carried by users in their mobile devices. These descriptions are used to select and compose networked services. Such a composition integrates services that are equivalent to the users task. Additionally, this composition should result in an application that maximizes the Quality of Service (QoS) of the functionality delivered to the user.

A drawback to the IST Amigo framework is that workflows are directly mapped to available services without any mediating process and without abstract layers. This leads to a lack of flexibility and reusability of compound services. That is, applications cannot be developed during runtime operations by reusing predefined compositions of abstract services that are appropriated for specific workflows. Additionally, applications cannot be easily reconfigured by varying the composition of abstract services that better represent user goals.

7.4   Task-Oriented Service Framework for Ubiquitous Computing

7.4.1   Task and Action Definition

A task is a representation of user goals that consists of a set of coordinated *actions*. To fulfill the translation of user goals represented in the *task layer* into appropriated services available in a Ubicomp environment, or on the Web, a task is decomposed into a set of actions. Thus, an action is a high-level function defined according to the users' perspective. The coordination of these actions is represented in advance during the design stage by service developers. Such a representation corresponds to a determined and descriptive method to organize actions either sequentially, iteratively or in parallel. This is known as an *application template* (see Fig. 7.2).

The *composition-pattern layer* supports the application templates of the task layer by mapping its actions to Web-based services and services instances that are tightly bound to a Ubicomp environment. This mapping process is performed through the use of reusable SCPs. These SCPs consist of a set of coordinated and reusable *abstract services* or functions that perform the actions of an application template (see Fig. 7.2). These SCPs are also created in advance by service developers during the design stage.

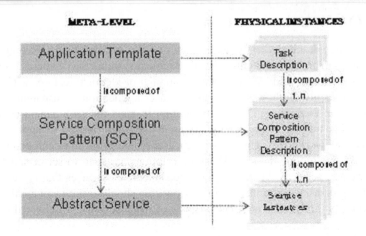

Fig. 7.2 Templates and Patterns for a User-centric Service Provision in Ubicomp

7.4.2   Overall Architecture of the Framework

This section describes the overall architecture of the task-oriented service framework as well as its essential elements and their roles[1]. As shown in Fig. 7.3, the three major elements of the framework architecture are the *task broker*, the *SCP broker*, and the *repository system*. Regarding the task broker, its major function concerns selecting available tasks from the task ontology that is stored in the repositories based on the contextual information. Details of the ontology are provided in the next section. The task selection begins when the *property matchmaker* matches the context information against the task property values in the task ontology. After searching for available tasks, the task broker yields candidates of appropriate tasks by ranking them through on a semantically based approach. This ranking is performed by the *semantic measurer* module. Among the candidates, the task selector selects the most appropriate task instance - an application template - based on user preferences and on the ranking of the tasks.

The SCP broker searches for available SCPs from the *service composition-pattern ontology* based on a semantically based approach, similar to that of the task broker. It selects the most appropriate SCP from among candidate SCPs based on the contextual information. This process is supported by the *interpreter* , which obtains appropriate descriptions of application templates or service composition patterns interacting with the *description manager* . The interpreter also parses the SCP description that is selected and extracts a list of abstract services which should be performed for an action. The list of abstract services is transferred to the *service discovery* as input data. It also parses the execution logic of abstract services, which

---

[1] A demo application example that shows a concrete realization of the role of these elements is described in section 7.4.5.2.

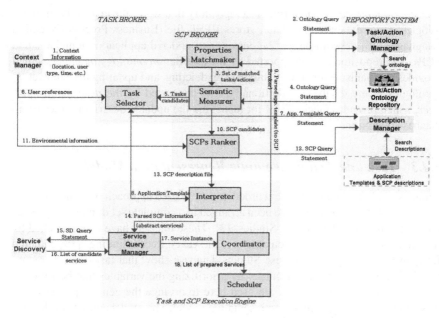

Fig. 7.3 Architecture of the Task-oriented Service Framework for Ubicomp

are coordinated in a sequential, iterative or parallel logic. It sends the parsed results of the application template to the SCP broker to select the most appropriate SCP, and sends the parsed results of the service composition patterns to the service query manager to search for available services. The SCP broker then formulates a query statement and send a service request to the *service query manager.*

The service query manager is an element that is dedicated to interacting with the service discovery module that searches for current services available on the Web or in a Ubicomp environment. It receives the query statement from the SCP broker and generates a string-formatted XML (Extensible Markup Language) query. This query is sent to the service discovery module and a list of available services is retrieved.

The *coordinator* binds the actual service instances retrieved from the service discovery module with the abstract services contained in the selected SCPs. After the service instances are bounded, the coordinator sends operations to services instances to execute and control them based on the coordination logic embedded in the SCP descriptions. The coordination is supported by the *service scheduler,* which prepares the actual execution of service instances as multiple requests for one service might occur. It negotiates conflicts that may take place between multiple requests of users or groups and forms SCPs collaborative by grouping or rescheduling services.

The *repository system* stores application templates and SCPs in distributed environments. In addition, it provides a mechanism to search for appropriate task descriptions and service composition patterns from distributed repositories. It is composed of the *task/action ontology repository* and the *application template/SCP description repository*. The former is a repository that stores the task and action on-tologies into a database. The latter stores BPEL file (Business Process Execution Language) [26] descriptions of the current and updated application templates and SCPs. The repositories are supported by the *task/action ontology manager*, which provides mechanisms for storing, modifying, deleting, and updating task and action ontologies in the database. Additionally, it is supported by the *description man-ager*, which provides mechanisms for managing the updated description files using database operations.

7.4.3   Task and Action Semantic Representation Model

Humans' daily tasks have been recorded in a structured approach over the past sev-eral decades. These records have been arranged in a complete set of publically avail-able datasets, termed Time-use Studies [3]. These datasets have been enriched by hundreds of thousands of participants over the years. Recent findings [19] gathered from an analysis of the Time-use Studies database show that tasks can be recog-nized with up to 80 percent accuracy by prioritizing the variables that best predict user activities. These variables are used here to arrange the generic properties of tasks and actions in a common representation model. The model consists of the task ontology, the action ontology and the SCP ontology as discussed in the previous section. These are shown in Fig. 7.4. Additionally, the current tasks contained in the Time-use Studies database are used to define a hierarchy of tasks and actions that are related to each other through subsumption [25] relationships (See Fig. 7.5). The semantic information embedded in these ontologies enables an enhancement in the selection of tasks when they are matched against the contextual information of users and environment status.

A task is generically described by a set of properties in which the values define a specific instance in the hierarchy. The *execution type* property determines if a task needs to be executed one time, iteratively or continuously. The *low-level context* property is composed of *time* and *location* . The former refers to the *day of the week* or *hour of the day*. The latter represents either a *symbolic location* - kitchen, bed-room, waiting room, corridor, yard, among other environments - or the coordinates (*latitude*, *longitude*) of a location where the task can be executed in the Ubicomp environment - for which a location system is required. A symbolic location can be also enriched with *social constraints* attached to a Ubicomp environment. They are omitted for simplicity in Fig. 7.4, as are the other primitive properties of *task name*, *synonym, user preference*, and *URI* (Uniform Resource Identifier). Physical infor-mation about the environment, such as the *brightness, level of noise, temperature* or *humidity*, is represented by the *environmental information* property. The informa-tion related to a user is embedded in the *user type* property . It consists of the *user status, age group, role* and *gender* of the user. A special property to determine if the task is collaborative, single or shared is included under the name *user interaction type*.

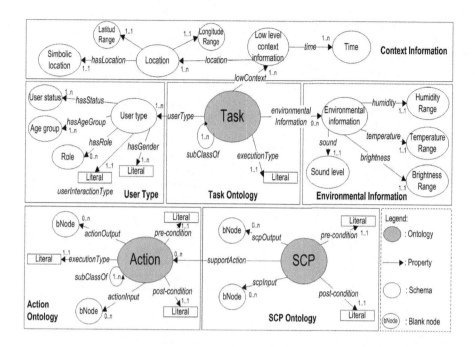

Fig. 7.4 Representation of Task, Action and SCP Semantics

Regarding the generic description of an action, the *pre-conditions* property states the necessary conditions that need to be met before an action can be instantiated - for instance, during a particular season, a specific hour of the day or after the execution of other action. On the other hand, having executed an action, this property may affect the Ubicomp environment by producing changes in the context or environmental conditions. These are represented through the *post-conditions* property . An action is also described by *inputs* and *outputs* . An input represents a message from a previous action, while an output is a message for the next action. A set of properties is not explicitly shown in Fig. 7.4 for simplicity. They are the *trigger* to activate an action, the *action name*, *URI*, and the *user type* and *environmental information* that are common for the ontologies in the task model. As shown in Fig. 7.4, an action is supported by one or more SCPs.

The properties that describe a task may be extended to reflect spatial, personal and social aspects [12]. In regards to spatial information the model may include properties like the *space constraints* (allowed level of noise, brightness or temperature), the *space level of public-ness* (public, quasi-public, private), the *space potential* (leisure, entertainment, business, etc.), among other properties. The personal aspect may be realized by including the *user preferences* (type of music, favorite movies, activities, food, etc), or the *relationship the user has with the space* (insider, owner, outsider, etc.), while the social aspect may state the *level of familiarity*

Fig. 7.5 Excerpt of a Task Ontology Example in the Home Domain

Fig. 7.6 Excerpt of an Action Ontology Example in the Home Domain

among a group of users (familiar, strangers, familiar strangers, etc). All these properties eventually may enrich the description of a task with regards to the space and its users.

The example of the model of tasks and actions introduced here is general enough to capture their generic properties. Fig. 7.5 and Fig. 7.6 show examples of the task and action ontology hierarchies for the home domain. For this domain, the task ontology may include metabolic, entertainment, sporting, personal care, duties, socializing, among other types of tasks.

7.4.4   Processes for Web Service Composition and Execution in Ubicomp

7.4.4.1   Overall Process for Dynamic Web Service Composition and Execution in Ubicomp

The overall process of dynamic Web service composition and execution making use of the proposed framework is as follows (see Fig. 7.7):

1. The task broker extracts task candidates from the task ontology based on the contextual information gathered from the context manager. To do this, the properties of task ontology are matched with the contextual information by the task broker.
2. The most appropriate task is selected by measuring the semantic distances between the ontology concepts and by considering the task ranking information.
3. The task execution engine extracts actions from the selected task description.
4. The SCP manager receives a parsed list of actions and context information from the task manager.
5. For each action, the SCP broker searches for SCP candidates from the action ontology that resides in the repositories.
6. The SCP broker selects the most appropriate SCP from amongst the candidates for each action by semantic matching similar to that used in the task selection and based on the contextual information. Abstract services in SCPs are composed together based on the execution logics in SCPs for executing actions.
7. Abstract service names, their operations and URIs, which are represented in each SCP, are sent to the service discovery, and find physical services that match the service description.
8. Service instances are dynamically bounded into composed abstract services through the service discovery routine.
9. Bounded service instances are executed by the service coordinator based on the execution logic to perform the actions in SCPs. Executions are performed action by action in sequential or parallel order.

Next subsections stress the task selection mechanism (steps 1, 2, and 3), and the SCP selection mechanism (steps 4, 5, and 6), related to the task broker and the SCP broker respectively. More details regarding these mechanisms, as well as concrete examples, can be found in [11, 12].

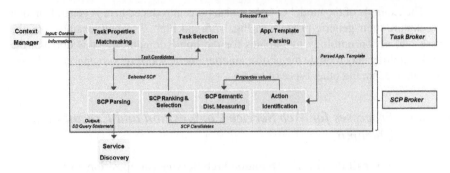

Fig. 7.7 Overall Process of Dynamic Service Composition

7.4.4.2   Task-Selection Mechanism

Regarding the properties that are best for predicting user tasks, Partridge et al. determined [19] that *location, hour of the day, day of the week* and *age group* are the variables with the highest contribution. However, it is surprising that jointly considering *location* with *hour of the day* on the one hand and *day of the week* with *age group* on the other hand is a more effective combination to recognizing user activities - reaching 80 percent accuracy on task recognition [19]. The task broker uses these two combinations of primary variables to select appropriate tasks in the task ontology . It matches the context information against the *location* and *hour of the day* properties in the first term, and *day of the week* and *age group in the second term*. The property matchmaking consists of counting the number of property values of the task ontology nodes that are completely matched against the values of the context information. This matchmaking generates a semantic matching value for the primary variables. The resulting set of task candidates is in turn matched against a second group of secondary variables that are less effective in recognizing user activities. This allows a reduction in the number of task candidates. These secondary properties correspond to (1) the *longitude* and *latitude* range; (2) the *users role, gender, interaction type* and *preferences*, and (3) all of the *environmental* properties. As with the primary variables, the matchmaking of the task ontology nodes against the secondary variables values generates a semantic matching value for the secondary variables.

The total semantic matching value is obtained for each selected task by weighting and adding the semantic matching value that result from the primary and secondary variables. The tasks are ranked based on the total semantic matching value. If two or more nodes share the same semantic matching value, a comparison based on their subsumption relationships is performed. This comparison is made possible given that the ontology is arranged in a hierarchy based on task functionalities. Additional technical detail concerning the task-selection mechanism is available in the literature [11].

7.4.4.3   Service Composition Pattern Selection Mechanism

After the task manager selects the most appropriate task in the task ontology , a task description is loaded as an instance of the task ontology. The task execution engine then parses the task description and creates a list of actions for performing the task. Each action in the description has its own URI, and the engine sends the URIs of actions to the service composition pattern broker. The SCP broker searches for available SCPs from the action ontology and extracts the candidate SCPs via property matchmaking first. A semantically based matching and ranking routine is then performed to select the most appropriate SCP from among candidates that best match the task selection. To select the most appropriate SCP from among the candidates, the SCP broker also matches the context information and user preferences with the values of the properties of action ontology. Additional detail regarding the selection of SCPs and the way in which they are bounded to available service instances is available in the literature [11]. Also available is a description of a task reconfiguration mechanism that addresses the dynamism of Ubicomp environments. The process of discovering service instances is described in another study [13].

7.4.5   Implementation and Evaluation

7.4.5.1   Implementation

The process of the creation of a Web service composition is included in the service framework for ubiquitous computing described in an earlier study [10], which is part of the Active Surroundings Middleware [9, 16]. The algorithms are implemented in Java with the following support: (1) Protege [30], a tool to create OWL [29] files with our task and action hierarchies; (2) Business Process Execution Language (BPEL) [26], a de facto language used to describe the coordination models of application templates and service composition patterns; (3) the Jena library [27], which is used here to implement semantic reasoning and assist with search queries from the loaded ontology model; (4) SPARQL [31], a query language and a protocol to access OWL [29] files by the W3C RDF Data Access Working Group , and (5) the My-SQL [28] database management system to manage ontologies, description files and Time-use Study datasets.

7.4.5.2   A Demo Application Example: The Ubiquitous DJ and the Follow Me TV

A demo was prepared in a test bed that shows the main elements of the proposed service framework for Ubicomp working together with other modules of the Active Surroundings Middleware [16]. The main purpose of this demo is to demonstrate the task selection and service composition mechanisms. Additionally, this demo attempts to illustrate how easy is to instantiate Ubicomp applications by making use of the framework. The mechanisms are shown by means of a demo in which users are supported with the services available at a simulated birthday party in the test bed. The demo is subdivided into three steps. The first step is the *watching TV* task at a location termed Kim's place; the second is a *dancing* task at Jae-Sung's place, and the third is a restart of the *watching TV* task but this time at Jae-Sung's place. The last step is termed the Follow Me TV application (see Fig. 7.8).

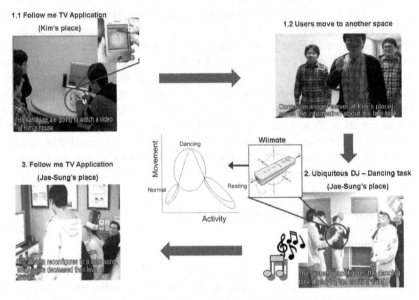

Fig. 7.8 Overall Concept of the Follow Me TV and Ubiquitous DJ Applications

The framework runs on a Windows XP platform with an Intel(R) Core(TM) 2 processor running at 1.87 GHz CPU with 2 GB of memory. The context manager and the service discovery protocol are distributed on different servers. The test bed is equipped with a real-time location sensor system [2] that provides the coordinates of each user in the room, which corresponds to the location *property*. In addition, the test bed is equipped to enable the context manager to generate input contextual information with reference to the *brightness*, *day of the week*, *hour of the day*, and several other contexts.

The following is a description of each step of this demo:

1. At the beginning, a group of users is engaged in the *watching TV* task at Kim's place. The user high-level context reported by the context manager indicates that users were *watching TV*, the *interaction type* was *collaborative* and the *symbolic space* was the *living room*. The service discovery found a suitable display and the context manager detected and stored the movie settings and user preferences. In order to enable the system to interoperate among multi-spaces,

---

[2] Ubisense, which is based on ultra-wideband (UWB) technology, which delivers 15cm 3D positional accuracy of objects in real time [32].

the context manager server at Kim's place saved the information about their last task when the users left Kim's place for the party.

2. At the party, each user carries a Wii remote - unofficially known as "Wiimote" - which is a remote controller for Nintendo's Wii console. In this case, Wiimotes are used to sense user motions. The context manager interprets each user's Wiimote raw motion data as the users level of activity, which is part of the input context information. In addition, the context manager aggregates the individual data to generate a group level of activity, another component of the input context information. The central graph in Fig. 7.8 shows that users have different levels of activity for different tasks, such as dancing or resting. Based on this group context, the context manager assigns the value *collaborative* to the *users' interaction type* property. When users moved to the dancing area, the system recognized the dancing task correctly. The essential input context information consists of (1) a *symbolic location (dancing area)*; (2) the *hour of the day (night)*; (3) the *day of the week (Saturday)* and (4) the *age group (Young)*. The secondary input context information consists of (1) the *location* (coordinates of the dancing area); (2) the *user interaction type (collaborative)* and (3) the *users level of activity (calm)*, which is mapped against the *user's status* property. Further details about privacy concerns of user information are out of the scope of this chapter. However, sensible data as the users age should be known at certain level, such as its range. By matching the above inputs against the task properties and narrowing down the total number of matched tasks, the algorithm selects the *dancing* task, and the result of which the application template contains the actions *retrieving favorite music content, deploying calm music* and *making the room darker*. Clearly the task selection itself would depend on how the task hierarchy is constructed. Such a construction may consider cultural factors, environmental constraints, and other concerns out of the scope of this work. The abstract services of the SCPs associated with the above actions confirm a service request and the service discovery then delivers the *audio search* service, the *audio service* and the *light control* service. However, the task was reconfigured when the level of activity increased to *normal*, selecting more active music. In fact, when users increase their motions, the level of activity becomes *normal* or *excited*, which does not match with the action *deploying calm music*. Therefore, the framework selects a new action by matching the new input with the action ontology properties and measuring the semantic distance between the candidate actions and the action *deploying calm music*. The algorithm then returns the actions *deploying normal/excited music* and *making the room lighter* for the *dancing* task.

3. When users changed their *symbolic location* at the party location from the *dancing area* to the *living room*, and when they stopped dancing, the system reconfigured and played a calm song. However, the context manager informed the system of the last task of which the *symbolic location* was Kim's place for this type of user. Hence, the system provides users with the option of executing the *watching TV* task. As the movie settings were stored as user preferences at Kim's place, the system deployed the same movie with the party location,

and the mellow song was turned off. We are currently extending our available services at the test bed, in order to allow the *dancing* task as well as the *watching TV* task to coexist in their execution. That is, in the case a group of users keep dancing, the movie may be played through personal devices. On the other hand, a recommendation service may suggest users to switch to another room to watch the movie through the device available there.

Table 7.1 Other Scenario Examples

Tasks	Actions
Having a breakfast	Making the room lighter; deploying calm music; deploying useful information - *showing e-mail, wheather forecast, e-newspapers front page, personal schedule*
Watching TV	Deploying TV content; controlling TV content access; recommending TV content
Exercising	Making sport schedule; monitoring health condition; monitoring sport schedule compliance; recommending exercises; recommening medication
Having a family party	Deploying music; deploying a background of family pictures; showing drinks mixing recommendations; guiding user for making drinks - *showing list of drinks, quatities of ingredients, recipes*; alarming specific locations - *alcohol and snack location*

Table 7.1 shows these actions in detail, including a number of other example task scenarios. For instance, the *exercising* task, a sporting task, can be selected receiving the context information consisting of *gymnasium, morning, week day, young*, the user *preferences on running*, the coordinates of the running machine, and a *single users interaction type*.

7.5 Summary

This chapter is concerned with using a task-oriented computing approach to realize the user centricity that characterizes Ubicomp environments. In these environments, user goals must be supported by transparently delivering Web-based services and available service instances. The chapter describes the essential requirements for service provisioning in Ubicomp environments - user centricity, context awareness, composability, and dynamicity. Moreover, it provides the state of the art of existing related work for a user-centric and task-based service provision scheme.

This chapter also leverages the task-oriented computing approach. It proposes a general semantically based model that captures the generic properties of tasks and actions. A task represents user goals as a coordinated set of actions in the form of an application template. The task and action model is leveraged by a task-selection mechanism and a service-composition mechanism that collectively bridge the gap between tasks and services. These mechanisms make use of three composition layers: the task layer, the composition-pattern layer , and the service layer. The task selection process is achieved by semantically matching the task property values against the contextual information of users and environment status. The result generated in the task layer is used in the composition-pattern layer by mapping the actions of a selected task to SCPs, which consist of a set of coordinated abstract services. This pattern-based approach for service composition ensures flexibility for developing Ubicomp applications and makes it easier to reconfigure tasks based on changes of user goals and other contextual information. The proposed solution provides reusability of application templates and of SCPs. The former can be reused for a specific task, and the latter for several action supports. Moreover, this solution provides dynamism through a dynamic composition of a SCP. Finally, a strategy is adopted that separates user concerns from the system perspective.

The authors are currently extending this task-oriented computing approach to compose tasks in an urban computing [14] domain, an extension of Ubicomp [12]. In this domain, it is necessary to support users with Web-based services and services in general. Those users consist of diverse social groups that are created from spontaneous interactions in parks, plazas, cafes, and other public spaces. Supporting these social groups requires the creation of tasks that are not defined in advance. Thus, an spontaneous task-composition mechanism that reacts to the context embedded in an urban locale is required [7, 23]. We are also working on task prediction while making use of Time-use Studies [3], extracting user behavioral patterns that can be used to prepare a task plan in advance.

Acknowledgements. This work was supported at the IT R&D program of MKE/KEIT under grant KI001877 [Locational/Societal Relation-Aware Social Media Service Technology]. The authors thank Jun-Sung Kim, Gonzalo Huerta-Canepa, Yong-Jae Lee, Byoungoh Kim, Jae-Sung Ku, and SaeHyong Park.

References

1. Dey, A., Salber, S., Futakawa, M., Abowd, G.: An Architecture to Support Context-Aware Applications. GVU Technical Report GIT-GVU-99-23 (1999)
2. Dey, A.: Understanding and Using Context. Personal and Ubiquitous Computing 5(1), 4–7 (2001)
3. American Time-use Study, http://www.bls.gov/tus/
4. Baldauf, M., Dustdar, S., Rosenberg, F.: A Survey on Context-Aware Systems. International Journal of Ad Hoc and Ubiquitous Computing (2004)
5. Biegel, G., Cahill, V.: A Framework for Developing Mobile, Context-Aware Applications. In: Proceedings of the 2nd IEEE Conference on Pervasive Computing and Communication (2004)
6. Bardram, J.E.: From Desktop Task Management to Ubiquitous Activity Based Computing. In: Integrated Digital Work Environments, pp. 49–78. MIT Press, Cambridge (2007)
7. Carmona, M., et al.: Public Places- Urban Spaces, The Dimensions of Urban Design. Architectural Press, London (2003)

8. Garlan, D., Siewiorek, D.P., Smailagic, A., Steenkiste, P.: Project Aura: Toward Distraction-free Pervasive Computing. In: IEEE Pervasive Computing (April-June 2002)
9. Huerta-Canepa, G.F., Jimenez-Molina, A.A., Ko, I.Y., Lee, D.: Adaptive Activity based Middleware. IEEE Pervasive Computing 7(2), 58–61 (2008)
10. Jimenez-Molina, A.A., Koo, H., Ko, I.Y.: A Template-Based Mechanism for Dynamic Service Composition Based on Context Prediction in Ubicomp Applications. In: International Workshop on Intelligent Web Based Tools. IEEE ICTAI 2007 (2007)
11. Jimenez-Molina, A.A., Kim, J., Koo, H., Ko, I.: A Semantically-based Task Model and Selection Mechanism in Ubiquitous Computing Environments. In: Velásquez, J.D., Ríos, S.A., Howlett, R.J., Jain, L.C. (eds.) Knowledge-Based and Intelligent Information and Engineering Systems. LNCS, vol. 5711. Springer, Heidelberg (2009)
12. Jimenez-Molina, A., Kang, B., Kim, J., Ko, I.Y.: A Task-oriented Approach to Support Spontaneous Interactions among Users in Urban Computing Environments. In: Proceedings of the 8th IEEE International Conference on Pervasive Computing and Communications (PerCom 2010). 7th International Workshop on Managing Ubiquitous Communications and Services (MUCS 2010), Mannheim, Germany (to appear, 2010) (Accepted)
13. Kang, S., Kim, W., Lee, D., Lee, Y.H.: Group Context-aware Service Discovery for Supporting Continuous Service Availability. In: 3rd International Workshop on Personalized Context Modeling and Management for UbiComp Applications (2005)
14. Kindberg, T., Chalmers, M., Paulos, E.: Guest Editors' Introduction: Urban Computing. IEEE Pervasive Computing 6(3) (July-September 2007)
15. Ko, I.: User-centric and Intelligent Service Composition in Ubiquitous Computing Environments. In: Velásquez, J.D., Ríos, S.A., Howlett, R.J., Jain, L.C. (eds.) Knowledge-Based and Intelligent Information and Engineering Systems. LNCS, vol. 5711, p. 375. Springer, Heidelberg (2009)
16. Lee, D.: Active Surroundings: A Group-aware Middleware for Embedded Application Systems. In: 28th Intl. Conference on Computer Software and Applications, pp. 404–405. IEEE, Los Alamitos (2004)
17. Mokhtar, S.B., Liu, J., Georgantas, N., Issarny, V.: QoS-aware Dynamic Service Composition in Ambient Intelligence Environments. In: 20th IEEE/ACM International Conference on Automated Software Engineering, pp. 317–320. ACM, New York (2005)
18. Moran, T., Dourish, P.: Introduction to Context-Aware Computing. IBM Almaden Research Center, University of California, Irvine. Special Issue of Human Computer Interaction 16 (2001)
19. Partridge, K., Golle, P.: On Using Existing Time-Use Study Data for Ubiquitous Computing Applications. In: 10th Intl. Conference on Ubiquitous Computing, pp. 144–153. ACM, New York (2008)
20. Ranganathan, A., Chetan, S., Campbell, R.: Mobile Polymorphic Applications in Ubiquitous Computing Environments. In: Mobile and Ubiquitous Systems: Networking and Services, Mobiquitous (2004)
21. Roman, M., Hess, C.K., Cerqueira, R., Ranganathan, A., Campbell, R.H., Nahrstedt, K.: Gaia: A Middleware Infrastructure to Enable Active Spaces. In: IEEE Pervasive Computing, pp. 74–83. IEEE, Los Alamitos (2002)
22. Satyanarayanan, M.: Pervasive computing: vision and challenges. In: IEEE Personal Communications, vol. 8. Carnegie Mellon Univ., Pittsburgh (August 2001)
23. Schieck, A.F., Briones, C., Mottram, C.: A Sense of Place and Pervasive Computing Within the Urban Landscape. In: 6th International Space Syntax Symposium (2007)
24. Sousa, J.P., Garlan, D.: Aura: an Architectural Framework for User Mobility in Ubiquitous Computing Environments. In: 3rd Working IEEE/IFIP Conf on Software Architecture (2002)

25. Definition of Subsumption, http://www.thefreedictionary.com/subsumption
26. The Business Process Execution Language,
    http://docs.oasis-open.org/wsbpel/2.0/wsbpel-v2.0.pdf
27. The Jena Semantic Web Framework, http://jena.sourceforge.net
28. The MySQL Data Base Management System, http://www.mysql.com
29. The OWL Web Ontology Language, http://www.w3.org/TR/owl-features
30. The Protege Ontology Editor and Knowledge-base Framework,
    http://protege.stanford.edu
31. The SPARQL query language for RDF,
    http://www.w3.org/2009/sparql/wiki/MainPage
32. Ubisense Real-time Location System, http://www.ubisense.net
33. Wang, Z., Garlan, D.: Task-Driven Computing. Technical Report, CMU - CS -00-154
    (2000)
34. Weiser, M.: The Computer for the 21st Century. In: Scientific American. Reprinted in
    IEEE Pervasive Computing, pp. 94–104. IEEE, Los Alamitos (2003)

Chapter 8
Ontological Engineering and the Semantic Web

José Manuel Gómez-Pérez and Carlos Ruiz

Abstract. This chapter focuses on ontologies as means of representing knowledge and reason across the various domains. In this chapter, the state of the art in ontology engineering from a number of different perspectives is reviewed ranging from methodologies to manage the ontology lifecycle, from their inception to maintenance, to knowledge representing languages, reasoning methods and technology, and ontology development frameworks. The role of *networked ontologies* as a new stage in ontology engineering is described, where we transit from isolated information silos to networks of knowledge. Key issues of networked ontologies like modularization, customization and ontology alignment are approached. The different methods, technologies, and frameworks described herein with real life applications using ontologies are illustrated. This chapter is concluded with the introduction of key concepts for future work in ontology engineering and the Semantic Web.

8.1 Introduction to Knowledge Representation and Ontology Engineering

Ontological Engineering refers to the set of activities that concern the ontology development process, the ontology life cycle, the methods and methodologies for building ontologies, and the tool suites and languages that support them. During the last years, increasing attention has been focused on ontologies and Ontological Engineering. Ontologies are now widely used in Knowledge Engineering, Artificial Intelligence and Computer Science; in applications related to knowledge management, natural language processing, e-commerce, intelligent integration information, information retrieval, integration of databases, bioinformatics, and education; and in emerging fields such as the Semantic Web. This chapter aims to acquaint readers

José Manuel Gómez-Pérez · Carlos Ruiz
iSOCO, Intelligent Software Components S.A
e-mail: jmgomez@isoco.com, cruiz@isoco.com

J.D. Velásquez and L.C. Jain (Eds.): Advanced Techniques in Web Intelligence – 1, SCI 311, pp. 191–224.
springerlink.com   © Springer-Verlag Berlin Heidelberg 2010

with the basic concepts and major issues of ontologies and especially networked ontologies, making them more understandable to computer scientists that use ontologies in their information systems.

The adoption of the modelling paradigm by DARPA's Knowledge Sharing Effort [39] initiative allowed to envisage a way of building intelligent systems, based on assembling existing knowledge components rather than constructing knowledge bases from scratch. The reusability of available knowledge resources would relieve system developers from large part of the knowledge acquisition work, allowing them to focus on the creation of the knowledge resources and devices required to solve the specific parts of their systems. According to this approach, declarative, static domain knowledge is modelled by means of ontologies

Among the many definitions of ontologies, Gruber's [22] is the most quoted in literature: *An ontology is an explicit specification of a conceptualization.* This definition was later extended [7] and explained by Studer and colleagues [48]: *An ontology is a formal, explicit specification of a shared conceptualization. Which refers to an abstract model of some phenomenon in the world by having identified the relevant concepts of that phenomenon. Explicit means that the type of concepts used, and the constraints on their use are explicitly defined. Formal refers to the fact that the ontology should be machine-readable. Shared reflects the notion that an ontology captures consensual knowledge, that is, it is not the private life of some individual, but accepted by a group.*

Guarino [23] identifies the main benefits of using ontologies as enablers of a 'higher' level of reuse with respect to the usual case in software engineering (knowledge reuse instead of software reuse). Ontologies enable developers to reuse and share application domain knowledge using a common vocabulary across heterogeneous software platforms, eventually allowing them to concentrate on the structure of the domain and the task at hand, and therefore protecting them from being bothered by implementation details.

The emergence of the Semantic Web has marked a milestone in the evolution of ontologies. According to Berners-Lee [5], the Semantic Web is an extension of the current Web in which information is given well-defined meaning, better enabling computers and people to work in cooperation. This cooperation can be achieved by using shared knowledge-components, and so ontologies have become key instruments in developing the Semantic Web.

The growing availability of information has shifted the focus from closed, relatively data-poor applications, to mechanisms and applications for searching, integrating and making use of the vast amounts of information that are now available. Ontologies provide the semantic underpinning enabling intelligent access, integration, sharing and use of data and indeed this technology has now become strategic. Ontologies provide a key technology to support interoperability on the web and for enabling semantic integration of both data and processes. Indeed, ontologies have been so successful in recent years, that even 'traditional' companies, such as IBM, have now released their own ontology management systems, while market research firms, such as Gartner, rank taxonomies/ontologies third in their list of the top 10 technologies in recent years. Hence, a decade after the notion of ontology was first proposed by Gruber, ontologies are coming of age, and we are now entering a new phase, one in which ontologies are being produced in larger numbers and exhibit greater complexity than ever before. Thus, we are now facing both new opportunities and new challenges.

As ontologies are produced in larger numbers and exhibit greater complexity and scale, we now have an opportunity to build a new generation of complex systems, which can make the most of the unprecedented availability of both large volumes of data and large, reusable semantic resources. These systems will provide new functionalities in the emerging semantic web , in the automation of business to business relationships, and also in company intranets. At the same time, we face a challenge: current methodologies and technologies, inherited from the days of closed, data-poor systems, are simply not adequate to support the whole application development lifecycle for this new class of semantic applications.

The aim of projects like NeOn is to create the required infrastructure and methodology to support the development life-cycle of such a new generation of semantic applications, with the overall goal of extending the state of the art with economically viable solutions. These applications will rely on a network of contextualized ontologies, exhibiting local but not necessarily global consistency.

This chapter echoes the overall aim of project NeOn and provides a survey on ontologies as means to represent knowledge and reason with it across the various domains. The chapter is as follows:

- Section 8.2 reviews the state of the art in methodologies for ontology engineering to manage the ontology lifecycle from their inception to their maintenance.
- Section 8.3 describes reasoning technology and methods; Section 8.4 outlines techniques like modularization and customization.
- Section 8.5 emphasizes the role of *networked ontologies* as a new stage in ontology engineering, where we span from isolated information silos to networks of knowledge.
- Section 8.6 approaches key issues of networked ontologies like context and ontology alignment also it describes the required frameworks for ontology development.
- Finally, Section 8.7 illustrates the different methods, technologies, and frameworks with real life applications using ontologies. Section 8.8 concludes this chapter with the introduction of key concepts for future work in ontology engineering and the Semantic Web: Linked Data and Provenance.

8.2   A Methodological Approach to Ontology Engineering

This section focuses on how to manage the lifecycle and development processes associated with ontologies i.e. methodological approaches to ontology engineering. Research on methodologies for the Ontology Engineering field is still in its "adolescent stage"'. The 1990s and the first years of this new millennium have witnessed a growing interest in many practitioners regarding approaches that support the creation and management as well as the population of single ontologies built from scratch. Such methodological approaches (e.g., METHONTOLOGY [16], On-To-Knowledge [46], and DILIGENT [44] are slowly turning the art of constructing ontologies into an engineering activity.

Knowledge engineers need to capture, represent, analyse and exploit knowledge in order to produce a successful knowledge-based system. A methodological development of an ontology engineering project requires the definition and standardization of development and maintenance processes (from the requirements specification to the maintenance of the final product) and of the life cycle models to occur within the project. Based on the field of Software Engineering, methodologies for building knowledge-based systems (KBSs) have been proposed, taking into account the features of this kind of system that provide the complete life cycle for the ontology development process, including guidelines to be followed in the differing activities of the process.

One can differentiate between ontology development process and ontology life-cycle model, where the later establishes the order in which the activities performed in the former need to take place. In this section, we focus on METHONTOLOGY as the methodology from amongst those introduced above all used to describe and illustrate such concepts.

8.2.1 The Ontology Development Process according to METHONTOLOGY

The ontology development process defined in METHONTOLOGY refers to *which* activities are performed when building ontologies. It is crucial to identify these activities if agreement is to be reached on ontologies built by heterogeneous co-operative teams. Such activities are organized into the three following categories: management, development and support.

1. **Ontology management activities** include activities that initiate, monitor, and control an ontology project throughout its life cycle explained as follows:

   - The *scheduling* activity identifies the tasks to be performed, their arrangement, and the time and resources needed for their completion. This activity is essential for ontologies that use ontologies stored in ontology libraries or for ontologies that require a high level of abstraction and generality.
   - The *control* activity guarantees that scheduled tasks are completed in the manner intended to be performed.
   - The *quality assurance* activity assures that the quality of each and every product output (ontology, software and documentation) is satisfactory.

2. **Ontology development oriented activities** are grouped into pre-development, development and post-development activities. The **pre-development activities** include activities that explore and allocate requirements before ontology development can begin. These activities are:

- An environment study is carried out to know the platforms where the ontology will be used, the applications where the ontology will be integrated, etc.
- A feasibility study answers questions such as: it possible to build the ontology? Is it suitable to build it?, etc.

The **development activities**include activities performed during the development and enhancement of an ontology project. Such activities are:

- The specification activity states why the ontology is being built, what its intended uses are and who the end-users are.
- The conceptualization activity structures the domain knowledge as meaningful models at the knowledge level [40].

The **post-development activities** include the maintenance activity, that updates and corrects the ontology if needed, and the (re)use activity, that refers to the (re)use of the ontology by other ontologies or applications.

3. **Ontology support activities** include activities that are necessary to assure the successful completion of an ontology project. This group includes a series of activities performed at the same time as the development-oriented activities, without which the ontology could not be built. Such activities are:

- The *knowledge acquisition activity* whose goal is to acquire knowledge from experts of a given domain or through some kind of (semi)automatic process, which is called ontology learning.
- The evaluation activity makes a technical judgment of the ontologies, of their associated software environments, and of the documentation. This judgment is made with respect to a frame of reference during each stage and between stages of the ontologys life cycle.
- The *integration activity* is required when building a new ontology by reusing other ontologies already available.
- The *merging activity* consists of obtaining a new ontology starting from several ontologies on the same domain. The resulting ontology is able to unify concepts, terminology, definitions, constraints, etc., from all the source ontologies. The merge of two or more ontologies can be carried out either in run-time or design time.
- The *alignment activity* establishes different kinds of mappings (or links) between the ontologies involved. Hence this option preserves the original ontologies and does not merge them.
- The *documentation activity* details, clearly and exhaustively, each and every one of the completed stages and products generated.
- The *configuration management* activity records all the versions of the documentation and of the ontology code to control the changes.

8.2.2 The Ontology Lifecycle Model according to METHONTOLOGY

As mentioned above, the ontology development process does not identify the order in which the activities should be performed. This is the role of the ontology life cycle. The ontology life cycle identifies when the activities should be carried out i.e. it identifies the set of stages through which the ontology moves during its life time, describes what activities are to be performed in each stage and how the stages are related (relation of precedence, return, etc.). METHONTOLOGY proposes an ontology building life cycle based on evolving prototypes because it allows for adding, changing, and removing terms in each new version (prototype). For each prototype, METHONTOLOGY proposes to begin with the schedule activity that identifies the tasks that are performed, their arrangement, and the time and resources needed for their completion. After that, the ontology specification activity starts and at the same time several activities begin we carried out throughout the management (control and quality assurance) and support processes (knowledge acquisition, integration, evaluation, documentation, and configuration management). All these management and support activities are performed in parallel with the development activities (specification, conceptualization, formalization, implementation and maintenance) during the whole life cycle of the ontology.

Once the first prototype has been specified, the conceptual model is built within the ontology conceptualization activity. This is like assembling a jigsaw puzzle with the pieces supplied by the knowledge acquisition activity, which is completed during the conceptualization. Then the formalization and implementation activities are carried out. If some error is detected after any of these activities, one can return to any of the previous activities to make modifications or refinements.

Figure 8.1 shows the ontology life cycle proposed in METHONTOLOGY, and summarizes the previous description. Note that the activities we carried out throughout the management and support processes are carried out simultaneously with the activities that we part of the development process.

Related to the support activities, Figure 8.1 also shows that the knowledge acquisition, integration and evaluation is greater during the ontology conceptualization, and that it decreases during formalization and implementation. The reasons for this greater effort are:

- Most of the knowledge is acquired at the beginning of the ontology construction.
- The integration of other ontologies into the one we are building is not postponed to the implementation activity. Before integration at the implementation level, integration at the knowledge level should be carried out.
- The ontology conceptualization must be evaluated accurately to avoid propagating errors in further stages of the ontology life cycle.

The relationships between the activities carried out during ontology development are called intra-dependencies, or equally, they define the ontology life cycle. METHONTOLOGY also considers that the activities performed during the development of an ontology may involve performing other activities in other ontologies already built

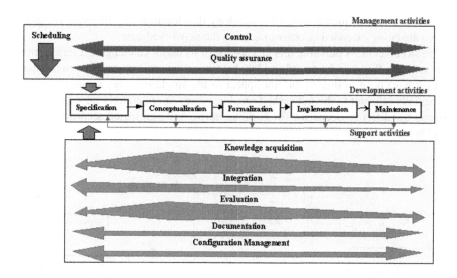

Fig. 8.1 The METHONTOLOGY ontology lifecycle

or under construction. Therefore, METHONTOLOGY considers not only intra-dependencies, but also inter-dependencies. Inter-dependencies are defined as the relationships between activities one carried out when building different ontologies. Instead of talking about the life cycle of an ontology, one should talk about crossed life cycles of ontologies. The reason is that, most of the times and before integrating an ontology in a new one, the ontology to be reused is modified or merged with other ontologies of the same domain.

8.2.3   Ongoing Work: Methodologies for Developing Networked Ontologies

As we will see in following sections, especially in section 8.5, ontologies are no longer being developed as isolated representations of particular domains. On the contrary, ontologies form networks of interlinked knowledge, which require new methodological approaches to be described properly. Indeed, there are two main issues related to the ontology network development process: i) the identification of when an ontology network is better than a single ontology for a particular case and ii) the impact of the evolution of components in a network, which is greater than in a single ontology. Hence, the development of ontology networks is a more complex process, which has some specific features different from those of building single ontologies. The main goal of the ontology network development process is to identify and define which activities are carried out when ontology networks are collaboratively built. Methodologies like METHONTOLOGY, used above to illustrate a typical methodology for ontology development and lifecycle management are being extended in order to address the networked case.

Project NeOn[1] is pursuing this task as one of its major challenges. A description of such an effort can be found in [34], here we enumerate some of the activities, which have been evaluated did not appear in the single ontology scenario but are especially relevant to the networked ontology cases or are entirely new. Figure 8.2 shows the different scenarios for building networked ontologies according to the NeOn methodology.

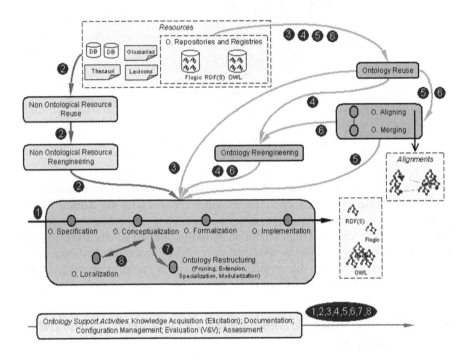

Fig. 8.2 The NeOn methodology lifecycle

- **Ontology Alignment** refers to the activity of finding the correspondences between two or more ontologies and storing/exploiting them.
- **Ontology Merging** refers to the activity of creating a new ontology or ontology module from two or more, possibly overlapping, source ontologies or ontology modules.
- **Ontology Customization** refers to the activity of adapting an ontology to a specific user's needs.
- **Ontology Integration** refers to the activity of including one ontology in another ontology.

---

[1] http://www.neon-project.org

- **Ontology Modularization** refers to the activity of identifying one or more modules in an ontology with the purpose of supporting reuse or maintenance.

8.3 Reasoning

A reasoner or semantic reasoner is a software tool that allows one to infer logical consequences from a set of asserted facts or axioms. The notion of semantic reasoning generalizes the notion of inference engine by providing a richer range of mechanisms and elements, besides, by defining inference rules through an ontology language and a language for universe description. Some reasoners use different types of logics for reasoning including first order logic and description logic (DL).

Focusing on the latter, OWL [4, 25] has features from several families of representation languages, including description logics. OWL was primarily designed to represent information about categories of objects and how objects are interrelated, characterised by the use of various constructors to build complex classes from simpler ones. Using these semantics, inferences about ontologies and individuals can be made. Besides, OWL emphasizes the decidability of key reasoning problems, and by the provision of sound, complete and (empirically) tractable reasoning services.

In addittion to the ability to describe concepts formally and answer questions about the concepts and instances described, the basic inference problems are:

- Subsumption: check whether a concept is a subset of another concept and, then, knowledge is correct
- Equivalence: check whether knowledge is minimally redundant
- Consistency: called satisfiability as well, check whether knowledge is meaningful and no contradiction among the definition exists.
- Instantiation: check whether individual i is instance of class C.

Around the aforementioned basic inference problems, research has focused on solutions for satisfiability: in particular, TBox and ABox reasoning, and tableau (sometimes referred as tableaux)procedures. On the one hand, while TBox statements validates an ontology in terms of the vocabulary used (a set of classes and properties), ABox are TBox-compliant statements about that vocabulary (e.g. facts associated with the ontology and the knowledge base). On the other hand, a semantic tableau [2] is a decision procedure to determine the satisfiability of an ontology using description logics.

In the following, information was provided about the main existing semantic reasoners:

- **Pellet**[2] is a Java-based reasoner for OWL2 from Clark and Parsia. It implements a tableau-based decision procedure for general TBoxes (subsumption, satisfiability, classification) and ABoxes (retrieval, query answering). It includes support for OWL 2 profiles including OWL 2 EL.

---

[2] http://clarkparsia.com/pellet/

- **OWLIM**[3] is semantic repository and reasoner, packaged as a Storage and Inference Layer (SAIL) for the Sesame RDF database from Ontoprise. OWLIM uses the TRREE engine -a native RDF database and rule-entailment engine- to perform RDFS, and OWL DLP reasoning. It performs forward-chaining of entailment rules on top of RDF graphs and employs a reasoning strategy. OWLIM offers configurable reasoning support and performance through rule-set definition and selection. OWLIM offers to flavors: (1) SwiftOWLIM, performing reasoning and query evaluation in-memory; and (2) BigOWLIM, operating with file-based indices, which allows it to scale to billions of statements.
- **FaCT++**[4] is a free software reasoner implemented in C++ in order to create a more efficient software tool and to maximise portability. It covers OWL and OWL 2 (lacks support for key constraints and some datatypes) DL-based ontology languages. It implements a tableau-based decision procedure for general TBoxes (subsumption, satisfiability, classification) and incomplete support of ABoxes (retrieval). It supports the Lisp-API and the DIG-API.
- **RacerPro**[5] is a commercial reasoner from Racer Systems developed in Lisp. The current version is RacerPro v2.0, which supports SHIQ and implements a tableau-based decision procedure for general TBoxes and ABoxes . It supports the OWL-API and the DIG-API and comes with numerous other features.

Table 8.1 provides an comparison overview of those reasoners, along with the development language, the ontology languages they support, their approaches to reasoning, support of rules, complexity, and availability.

Table 8.1 Comparison of different reasoners

Features	Pellet	OWLIM	FaCT++	RacerPro
Development language	Java	Java	C++	Lisp
Complexity supported	OWL-DL. SROIQ(D)	RDF	OWL-DL. SROIQ(D)	OWL-DL. SHIQ(D)
Reasoning Mechanism	Tableau	Forward-chaining on RDF graphs	Tableau	Tableau
Rule support	SWRL. DL Safe Rules	Own format	No	SWRL
Interface	DIG-API. Jena	-	Lisp-API. DIG-API	OWL-API. DIG-API
Licence	Dual	Dual	Open Source	Commercial
Organization	Clark and Parsia	Ontotext	U. Manchester	Racer Systems

---

[3] http://www.ontotext.com/owlim/
[4] http://owl.man.ac.uk/factplusplus/
[5] http://www.racer-systems.com/

8.4   Modularization and Costumization

Ontology modularization has attracted more and more attention in the recent years, as there is a clear need for such approaches to facilitate the management, maintenance and reuse, evolution and distribution of large, complex ontologies.

To some extent, the problem faced by ontology engineers can be seen as similar to the one faced by software engineers, where facilitating the management and reuser of the system requires the identification of components, modules, that can decoupled from this system, to be exploitable in a different context and integrated with different components. From the Semantic Web point of view, building an ontology as a combination of independent, reusable modules reduces the effort required for its management, in particular in a collaborative and distributed environment.

This topic corresponds to a variety of tasks including the design of modular ontologies and the modularization of existing ontologies. The following points describe the main activities within modularization:

Designing Modular Ontologies. The idea of ontology modularization is primarily inspired from the domain of software engineering where modularization refers to the design of software as the combination of self-contained components, easier to build, reuse and maintain than a program made of one, often large and intricate piece of code. Therefore, the most obvious scenario in which ontology modularization is involved is the case of the construction of an ontology not as a monolithic model, but taking benefiting from the properties of modular systems: reusability, extensibility and maintainability.

Partial Import and Reuse. Most of existing ontologies have not been designed with modularity in mind, hampering their integration and their reuse in applications other than the one they have been built for. Therefore, it is often required to be able to reuse and integrate only a part of a given ontology, by identifying the relevant elements within the ontology and by reusing them together with all the information required for their semantic definition. Ontology modularization techniques are needed to extract from potentially large and complex ontologies, modules that are relevant and adequate to the task at hand.

Improving Performance. Modularization enables one to focus only on the elements that are relevant for a given application at a given time having an impact for example, improving performance, by reducing the amount of knowledge that have to be manipulated by ontology-based tools, and by including reasoners and editors.

Modularization as a Means to Ontology Customization. A number of times a sub-set of ontological entities need to be treated differently than other parts of the ontology (e.g. due to confidentiality). Hence, it makes sense to replicate this situation and to create a number of partitions on the ontological model.

8.4.1   State of the Art

The following sections cover the state of the art in ontology modularization.

8.4.1.1 Knowledge Import

OWL Import. The OWL ontology language provides limited support to modular ontologies: an ontology document identified via its ontology URI can be imported by another document using the owl:imports statement. The semantics of this import statement is that all definitions contained in the imported ontology document are included in the importing ontology, as if they were defined in the importing ontology document. One of the most commonly mentioned weaknesses of the importing mechanism in OWL is that it does not provide any support for partial import [54, 43].

Partial/Semantic Import. Grau et al. [19] proposes a logic-based notion of modularity that allows the modeler to specify the external signature of their ontology, i.e. whose symbols that are reused from some an other ontology. The authors define two restrictions on the usage of the external signature, a syntactic and a slightly less restrictive semantic one, each of which is decidable and guarantees a certain kind of black-box behavior that enables the controlled merging of ontologies. To achieve this, certain constraints on the usage of the external signature need to be imposed: in particular, merging ontologies should be safe in the sense that they do not produce unexpected results such as new inconsistencies or subsumptions between imported symbols. A similar notion of partial import is introduced by Pan et al. [43].

Modularity in Classical Description Logics. Some authors [20] argue that modularity can be achieved without introducing new formal languages. In particular they propose translation of so called modular ontology languages (DDL, E-connections, etc.) into standard DL and describe new reasoning services to ensure modularity.

P-DL. Package-based Description Logics [3], use importing relations to connect local models. In contrast to OWL, which forces the model of an imported ontology to be completely embedded in a global model, the P-DL importing relation is partial in that only commonly shared terms are interpreted in the overlapping part of local models. Here, modules are called packages and subsets of the elements (entities) contained in a package can be imported by other packages.

8.4.1.2 Linking Modules

Distributed Description Logics. Distributed Description Logics (DDLs) [6] adopt a linking mechanism, relating the elements of *local ontologies* (called context) with elements of external ontologies (contexts). Each context is associated to its own local interpretation meanwhile semantic relations are used to correspond between elements of local interpretation domains expressed using bridge rules.

Context-OWL. C-OWL [8] derives from DDL by particularising the ontology language to OWL, and by enriching the family of bridge rules. However, the semantics of C-OWL is basically the same as the one of DDL but incorporating hybrid rules: a concept in an ontology is represented as a role in another ontology.

E-Connection. The E-connection approach [30, 21] helps to define link properties from one module to another. For example, if a moduleM1 contains a concept named 1:Fish and a moduleM2 contains a concept named 2:Region, one can connect these two modules by defining a link property named livesIn between 1:Fish and 2:Region.

Integrated Distributed Description Logics. IDDL [56] is another formalism for distributed reasoning upon networked DL knowledge base. Similarly to DDL, an IDDL interpretation allocates a different interpretation to each ontology but instead of relating domains directly, they are correlated in another domain called global domain of interpretation. The intuition behind this formalism is that ontology mappings may be provided by third party agents which activates correspondences from a point of view of both mapped ontologies. With that perspective in mind, using directional bridge rules like DDL is quite unsatisfactory.

8.4.1.3   Extracting Modules from Existing Ontologies

This section outlines techniques and tools that have been developed to help users extract or create modules from existing, and potentially large scale ontologies. Two major types of techniques can be distinguished: Ontology module extraction techniques and ontology partitioning. A more complete analysis of these tools has been carried out by d'Aquin el al. [10].

Ontology module extraction techniques. This task consists of creating a new module by reducing an ontology to the sub-part that covers a particular sub-vocabulary. This task has been called segmentation by Seidenberg and Rector [45] as well as traversal view extraction [41].

The traversal view extraction technique has been integrated in the PROMPT tool [41], to be used in the Protg environment. This approach recursively follows the properties around a selected class of the ontology, until a given distance is reached. The user can exclude certain properties in order to adapt the result to the needs of the application.

Another mechanism starts from a set of classes of the input ontology and extracts related elements on the basis of class subsumption and OWL restrictions [45]. Some optional filters can also be activated to reduce the size of the resulting module. This technique has been implemented to be used in the Galen project and relies on the Galen Upper Ontology.

Finally, Doran et al. [14] describes a technique focused on ontology module extraction for aiding an Ontology Engineer in reusing an ontology module. It takes a single class as input and extracts a module about this class. The approach it relies on is that, in most cases, elements that (directly or indirectly) make reference to the initial class should be included.

Ontology Partitioning. The task of partitioning an ontology is the process of splitting up the set of axioms into a set of modules $M_1,\dots,M_k$ such that each M_i is an ontology and the union of all modules is semantically equivalent to the original ontology O.

The approach by MacCartney et al. [35] aims at improving the efficiency of inference algorithms by localizing reasoning. For this purpose, this technique minimizes the shared language (i.e. the intersection of the signatures) of pairs of modules. A message passing algorithm to reason over the distributed ontology is proposed to implement resolution-based inference in the separate modules. Completeness and correctness of some resolution strategies is preserved and others trade completeness for efficiency.

A tool that produces sparsely connected modules of reduced size was presented by Stuckenschmidt and Klein [47]. The goal of this approach is to support maintenance and use of very large ontologies by providing the possibility of individually inspecting smaller parts of the ontology. The algorithm operates with a number of parameters that can be used to tune the result to the requirements of a given application.

8.4.1.4   Ontology Algebra

As modular ontologies are made of the combination of different ontology modules, operators are required to support the ontology designer in composing modules, creating them, and more generally, manipulating them. A few studies have been conducted on possible operators in an ontology algebra and, since an ontology module is essentially an ontology, these can be a source of inspiration for an ontology module algebra.

Wiederhold [55] defines a very simple ontology algebra, with the main purpose of facilitating ontology-based software composition. He defines a set of operators (Intersection, Union, and Difference) and applies a set-related operations on the entities described in the input ontologies, relying on equality mappings (=) between these entities.

In the same line of ideas, but in a more formalized and sophisticated way, Melnik et al. [36] describes a set of core operators (Match, Compose, Merge, Extract, Diff and Confluence) for model management, as defined in the RONDO platform [37]. The aim of model management is to facilitate and automatize the development of metadata-intensive applications by relying on the abstract and generic notion of model of the data, as well as on the idea of mappings between these models.

Kaushik et al. [29] define operators as a combination of ontologies created by different members of a community and written in RDF. This paper first provides a formalization of RDF to describe set-related operators such as intersection, union and difference. It also adds other kind of operators, such as the quotient of two ontologies O1 and O2 (collapsing O2 into one entity and pointing all the properties of O1 to entities of O2 to this particular entity) and the product of two ontologies (inversely, extending the properties of from O1 to O2 to all the entities of O2).

8.5   Networked Ontologies

The development of ontologies along the years have revealed that there are different ways or possibilities of building ontologies. It is not premature to affirm that a new ontology development paradigm is starting, whose emphasis is on the reuse and possible subsequent reengineering of knowledge awareness resources, the collaborative and argumentative ontology development, and the building of ontology networks, as opposed to custom-building new ontologies from scratch. Thus, an ontology network or a network of ontologies is defined as a collection of ontologies (called networked ontologies) [24] associated with a variety of different relationships such as mapping, modularization, version, and dependency relationships. This section outlines these different types of relationships.

8.5.1   Ontology Mapping

Ontology mapping (also ontology aligning) refers to the activity of finding the correspondences between two or more ontologies to obtain a more complete description and coverage of a domain, creating a net of knowledge without the need of creating bigger and bigger ontologies. Ontology Mapping can help to add additional context information to the ontologies, but doing it manually is a rather time-consuming process. This process is also referred as Ontology Aligning, while Ontology Matching can be seen as the first step of this process. Some existing ontology mapping algorithms have been outlined which execute the mapping automatically:

- Aroma [11] is an extensional matcher using linguistic and statistical techniques developed by INRIA. Aroma uses association rule mining and thus returns subsumption relationships.
- ASMOV [26] is a terminological, structural and extensional matcher developed by Infotech, USA. The result of individual matchers are aggregated by a weighted sum and ASMOV uses a semantic validator for feeding back the system with consequences and inconsistency detection which lead to adding and suppressing correspondences respectively.
- DSSim [38] is a multi-matcher system using Depster-Shafer theory for aggregating the results of each matcher. It is developed by Open University. Many of these matchers are based on the terminological part of the ontologies.
- Lily [32] is a combination of three matchers which are dynamically chosen with regard to the ontology to match. In particular when ontologies are large, they are first partitioned, then passed on to different terminological and structural matchers. Finally, a semantic matcher is able to postprocess the obtained alignments to filter correspondences. Lili is developed at Tsinghua University.
- RiMOM [50] is a multi strategy matcher which dynamically selects the techniques to be used on the basis of ontology characteristics. Most of the techniques are terminological or structural. When several strategies are used, their results are aggregated in a weighted average. RiMOM is also from Tsinghua University.
- SAMBO [31] is mostly a thesaurus-assisted string matching system developed at Linkping University. Like the other systems it also aggregates several classical techniques, e.g., structural.

8.5.2 Collaboration

This section presents overviews of the literature on collaboration in ontology engineering tools.

Initial collaborative features were included by first generation ontology tools that were used in the 90s. The most representative ones here outlined: Ontolingua [15] was designed specifically with collaboration in mind, supporting full synchronous editing of ontologies accessed by multiple users employing a standard web-browser; WebOnto [13] is based on a feature called broadcast and receive mode: a user enables the broadcast mode and then enters the edit mode, changes are broadcasted to other users in receiving mode (and where editing is not possible); In Ontosaurus [49] it is possible to work synchronously on an ontology, but every time modifications carried out the ontology is locked. Other users can see which user is editing and when the lock was created. After changes are made, the ontology must be updated and the user should return to the browse mode. This then removes the edit locks.

Regarding the current ontology editors, Protg offers Collaborative Protg extending the ontology environment with collaborative functionalities. In Collaborative Protg4 users can annotate ontology elements as well as ontology changes. Also voting and user interaction via chat are supported. Collaborative Protg has been developed as an extension of the multi-user Protg system, which already allowed multi-user-access and editing [52]. It works by having one dedicated server running that coordinates all users connecting to the server in client mode.

Moreover, some other semantic tools offer several collaborative features:

- OntoWiki [1], mainly a knowledge acquisition tool, is a purely web-based application that allows collaborative building of ontologies based on different views on the data. The system keeps track of all the changes applied to the knowledge base and users can review this information. When editing classes or properties, OntoWiki redirects to pOWL, a web-based ontology authoring and management tool.
- DBin [53] is a P2P system that allows users to collaboratively create knowledge bases. To facilitate this task, DBin allows the creation of domain- or task-specific user interfaces from a collection of components (such as ontology editors, views, forums or other plugins). Users can edit an ontology directly on the server and also track the changes made by other users. Provenance information is also provided by the system. DBin runs locally and connects to a P2P network from there.
- SWOOP [28] allows collaborative ontology development by using an Annotea [27] plugin. The generated annotations can then be published and distributed using any public Annotea Server.

Within the NeOn toolkit, the Cicero tool [12] facilitates an asynchronous discussion and decision taking process between participants of an ontology engineering project. It enhances efficiency supporting users in discussing the design rationale of ontologies recording the pro and contra arguments and leading to fewer redundancies in disputes. This facilitates efficient discussions and accelerates convergence to a solution. Beside, the captured discussions reflect the design rationale of an ontology. By attaching a discussion to the entities in the ontology, it is possible later to understand why certain elements are modeled as such. Furthermore, prior discussions can easily be resumed if e. g. new requirements have to be taken into account.

8.6   Ontology Development Frameworks

This section provides a description of the current state of ontology development tools, describing the core features and providing a comparison based on some functional and non-functional dimensions.

Ontology development tools and frameworks support the creation, edition and development of ontologies. These editors are designed to assist and facilitate the ontology development cycle, enabling the expression of ontologies in a range of languages languages, with different kinds of representation and views, and including plugins to allow reasoning, collaborative work, recommendation, etc..

The major players are:

Protégé. Protégé has been created by the Stanford Medical Informatics group at Stanford University.

Protégé 3.x is a Java-based open source standalone application to be installed and run on a local computer. It enables users to load and save OWL and RDF ontologies, edit and visualize classes, properties and SWRL rules, define logical class characteristics as OWL expressions and edit OWL individuals. With respect to the supported languages Protégé is a hybrid tool. The internal storage format of Protégé is frame-based. Therefore Protégé has native frame-support. The support for OWL is provided by a special plug-in that fits into the Protégé plug-in architecture. The Protégé-OWL API is built on top of the frame-based persistence API using frame-stores. The API provides classes and methods to load and save OWL files, to query and manipulate OWL data models, and to perform reasoning based on Description Logic engines. The API is designed to be used in two contexts: (1) development of components that are executed inside the Protégé UI, and (2) development of stand-alone applications (e.g., Swing applications, Servlets or Eclipse plug-ins).

Protégé 4.0 has been in development in parallel with Protégé 3.x. Protégé 4 is completely new reimplementation: It reimplements a familiar interface on top of the OWL API (previously known as Wonderweb API). Initial alpha versions were developed by the COODE team and it is now under active development by CO-ODE, Stanford's Protégé team, and several other developers.

Top Braid. TopBraid Composer is a modelling tool for the creation and maintenance of ontologies. It is a complete editor for RDF(S) and OWL models. TopBraid Composer is built upon the Eclipse platform and uses Jena as its underlying API. The system has the open source DL reasoner Pellet built-in as its default inference engine, but other classifiers can be accessed via the DIG interface. Historically the development of TopBraid Composer has its roots in Protégé OWL. Thus some of the concepts of TopBraid are similar to those of Protégé, such as the generation of schema-based forms for data acquisition. The most obvious difference from a technical perspective is the usage of the Eclipse platform as a base and the lack of the Frame-based.

SWOOP. SWOOP is an open-source hypermedia-based OWL ontology editor, originally developed at the University of Maryland. The user interface design of SWOOP follows a browser paradigm, including the typical navigation features like history buttons. Offering an environment with a look and feel known from web-browsers, the developers of SWOOP aimed at a concept that average users are expected to accept within a short time. Thus users are enabled to view and edit OWL-ontologies in a web-like manner, which concerns the navigation via hyperlinks but also annotation features. SWOOP therefore provides an alternative to web-based ontology tools but offers additional features such as a plug-in mechanism. SWOOP is designed as a native OWL-editor, which supports multiple OWL ontologies and consistency checking based on the capabilities of attached reasoners. Core features of SWOOP are the debugging features for OWL ontologies, exploiting features of OWL reasoners (in this case Pellet). This includes for example the automatic generation of explanations for a set of unsatisfiable axioms (e.g. for a particular class).

The NeOn Toolkit. The NeOn toolkit is a state-of-the-art, open source multi-platform ontology engineering environment, which aims to provide comprehensive support for all activities in the ontology engineering life-cycle. Basic ontology management and editing functionalities are provided by the core NeOn Toolkit. Plugins extend the core NeOn Toolkit with additional functionalities supporting specific life-cycle activities via the update-site mechanism.

The core contains plugins for ontology management and visualization. The core features include:

- Basic Editing of F-Logic and OWL DL ontologies
- Visualization/Browsing
- Import/Export/Transformations between F-Logic, RDF(S), and OWL
- Queries and Reasoners

A number of plugins extends the toolkit by various functionalities, including:

- Rule Support: Graphical/Textual editing, debugging
- Mediation: Graphical Mapping Editor, life-interpretation of mappings
- Database Integration: Database schema import, database-access
- Queries: Query-Editor and persistent queries

The NeOn Toolkit relies on the architectural concepts of the Eclipse platform to enable the development of plugins: The Eclipse IDE (integrated development environment) provides both GUI level components as well as a plugin framework for providing extensions to the base platform. The toolkit provides an extensive set of plug-ins (currently 45 plug-ins are available) covering all aspects of ontology engineering, including: relational database integration, modularisation, visualisation, alignment, and project management.

The functionality of the NeOn Toolkit can be easily extended with powerful plugins in the following categories: Annotation and Documentation, Human-Ontology

Interaction, Modularization and Customization, Ontology Debugging, Ontology Dynamics, Ontology Evaluation, Ontology Matching, Ontology Specification, Reasoning and Inference, and Reuse.

8.6.1   Comparison among the Ontology Development Tools

For the evaluation among the different tools, a number of dimensions have been selected concerning both functional and non-functional aspects. The dimensions to objective measure were restricted and subjective impressions disregarded such as ease of use, coolness of the interface, etc.

Dimensions concerning non-functional aspects:

- License: What is the licensing model, what kind of free/commercial use does it allow
- Language and standard support: Which languages and standards are supported by the platform
- Open APIs: Whether the platform provides/uses open/standardized APIs
- Extensibility: Means for extending the platforms via plug-ins

Dimensions concerning functional aspects:

- Visualization: Graphic representation of model contents
- Reasoners: Integration with reasoners (internal or external), e.g. with Pellet or KAON2
- OWL support: Ability to support editing/reasoning/validating all OWL constructs
- Rule support: Ability to support editing/reasoning/validating rules
- Mapping support: Ability to support mappings
- Multiple ontology management: Ability to manage multiple ontologies
- Modular ontologies: Ability to manage ontologies in a modular way
- Querying: Support for querying ontologies, esp. at a higher level than SPARQL
- Integration with standard databases, UML, etc.
- Re-engineering: Re-using folksonomies, lexical resources, Formal concept analysis
- Multilinguality: Support for ontologies in multiple languages
- Design patterns: Support for ontology development from design patterns
- Collaboration: Support for collaborative development of ontologies, including integration with Wiki
- Support documentation/reporting/versions/etc.: A number of facilities, e.g., generating reports from ontology projects, versioning support, integration with documents, etc.
- Fine-grained annotation/ status tracking: Ability to annotate entities in an ontology with status information (broken, incomplete, redundant, obsolete, etc.)

The following table summarizes the main characteristics of the most important ontology development tools:

Table 8.2 Ontology development frameworks comparison. Non-functional features

Dimensions	Protégé 3	Protgé 4	TopBraid Composer	SWOOP	NeOn Toolkit
Non-Functional					
License	Mozilla public license, allows commercial use	Undefined, builds on several libraries under LGPL	Commercial license	MIT license, allows commercial use	Open-source Eclipse Public License
Languages	OWL 1 on top of Frames	OWL 2	OWL 1, partial support for OWL 2	OWL 1	OWL1 and 2
Open APIs	Proprietary OWL API on top of Frame API	OWL API	Jena API	OWL API	OWL API 3.0
Extensibility	Proprietary Plug-ins	Proprietary Plug-ins, using OSGi Framework	Eclipse Plug-ins	Proprietary Plug-ins	Eclipse Plug-ins

8.7 Applications

Imagine a world where cars autonomously attend to drivers requests according to the driving context, a world where citizens and public administrations speak the same language, a world where people can collaborate at work or home supported by virtual desktops which intelligently anticipate the services required by the user at each moment.

All these scenarios are becoming part of our daily lives, and semantic technologies are decisively contributing to it. On the other hand, the technology push including annotation, natural language processing, multimodality, context awareness, customization, and service and data integration is reaching the required level of maturity to satisfy this demand. All these factors converge in a new paradigm for work and leisure which represents new business opportunities for semantic technologies.

Out of available market studies and personal experience, latest and future trends in semantic technologies can be represented as a tridimensional space with three, intertwining, main axes (Figure 8.3): natural interaction, service-oriented architectures, and Web 2.0.

8.7.1 Natural Interaction: Focus on the User

Natural interaction is the most immediate field of application for semantic technology, aimed toward improving user experience during interaction with online information systems. In this context, natural language techniques are exploited to interpret user requests, combined with high precision and recall of query answers

Table 8.3 Ontology development frameworks comparison. Functional features

Dimensions	Protégé 3	Protégé 4	TopBraid Composer	SWOOP	NeOn Toolkit
Functional					
Visualization	Many visualizers. Most do not work well with OWL ontologies. Customizable forms	Cloud views. Most do on-views for editing/browsing. OWLViz. Label view	Matrix views. Hierarchical navigation. Integrated UML diagramming. Customizable forms. Label view	Browser-like navigation. Workflow editor. Supports	Tree, graph, relational, and hierarchical navigation. Summarize view.
Reasoners	Supports several reasoners via DIG	Pellet, Fact++ built-in, others via OWL API	several reasoners via DIG	Pellet built-in. Good debugging support	Pellet and Hermit
OWL Support	OWL 1 support	OWL 2 support. Several syntaxes and datatypes support	OWL 1. Several syntaxes	Support for axioms and many syntaxes	OWL2 Support. OWL API 3.0
Rule Support	SWRL extensions, JESS plug-in	-	SWRL editor, support for reasoning	-	Graphical Rule Editor and Debugger
Mapping Support	Prompt				
Multiple Ontology Mmnt.	Very limited	Manages multiple ontologies	Very good at multiple ontologies, consistency, other things	Manages multiple ontologies	Networked ontologies
Modular Ontologies	Regular OWL imports	Modular reuse based on syntactic locality	Regular OWL imports	Modular reuse based on syntactic locality	Plug-ins for modularization, reuse, merging, and alignment
Querying Standards	SPARQL. Several 3rd party plugins	DL-Querying. Under development	SPARQL	Enriched RQL	SPARQL. SAIQL
Re-engineering	OntoLing plug-in for reengineering from WordNet	Imports spreadsheet	Relational views via D2RQ. UML, XSD, and java classes import/export. Imports spreadsheet and xsd. RDFa and RSS feeds	-	Several plug-ins: gOntt, Oyster.
Linguistic support	OntoLing plug-in for lexicalizing from WordNet	-	-	-	Through plug-ins (e.g. Label-Translator)
Design patterns	Wizards for some logical patterns	Wizards for some logical patterns	-	-	Wizards. Ontology design portal, and eXtreme Design Tool
Collaboration	CollabProtégé	-	Limited creation of wiki pages out of ontology elements. Eclipse support for tasks and bookmarks	-	Cicero, Cupboard, Watson
Support for annotation and documentation	OWL doc production	OWL doc production	Best reporting tool	-	OWL doc production
Fine-grained annotation tracking	CollabProtg	-	-	Annotea and Change tracker	Change capturing, Cupboard

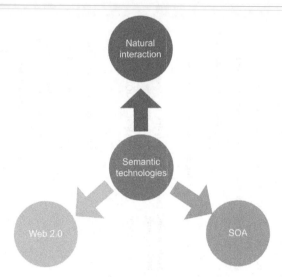

Fig. 8.3 Trends in Semantic Technologies

and multimodality allows users to move seamlessly between different modes of interaction.

There is an additional dimension in multimodality, customization, which provides users with the capability to adapt the system to better suit their goals. Context management plays a fundamental role here. Being aware of context, it is possible to maximize user satisfaction, producing more accurate answers to the specific circumstances in which the query was formulated.

Focused on user perception, natural interaction systems will be endowed with explanation features that will provide users with feedback on the actions accomplished by the system. It is at least as important to produce a comprehensive explanation of how a result was obtained as the correctness of the result itself.

Applications of semantic technology to natural interaction span across domains like eGovernment or the automotive industry. The use of semantic technology in this scenario allows two main goals to be achieved: i) citizens must not be forced to use administrative jargon, allowing them to naturally express themselves, and ii) citizens need to easily access to precise information that satisfies their queries.

In this regard, current developments belong to the first stage of the natural interaction roadmap (Figure 8.4), intensively using intelligent search engines where semantic technologies detect those services aimed by citizens in their queries, e.g. "I want to move to a new house" or "what guided tours can I do?". According to the roadmap, next steps will introduce mobility and voice interfaces in order to facilitate user interaction, including avatars that will improve user-friendliness. Finally, not only will services be recognized, but also executed. Thus, B2Bi will occur, involving third party services like e.g. banks and telcos.

In the automotive domain, context awareness is becoming a key issue. Current speech control systems require user commands to be expressed by the driver with

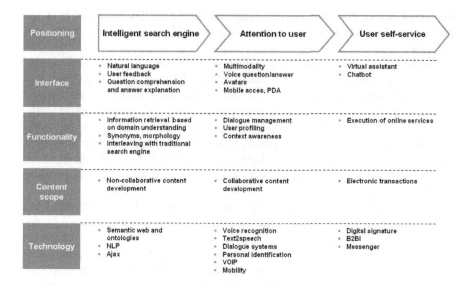

Fig. 8.4 Roadmap to Natural Interaction

high precision in order to obtain the right answer from the system, hence being a source for ambiguity and imprecise system responses. Additionally, there are other scenarios in which a context-aware speech dialog system helps the driver to deal with problems. For example, if the driver asks the navigation system to take her to a "good" restaurant, a context aware system can recommend a list of currently open, nearby restaurants which are aligned with the type of food e.g. mexican, italian, etc. that she usually prefers in accordance with previous experiences.

Driver commands can trigger different actions depending on the context. For example, it is not possible for regular speech recognition systems to extract a given command or display useful dialogs from questions like "what is this?". Without context information, this expression is meaningless for the car system. However, if context is appropriately formalized, it is possible to ground the question in the domain and provide a useful response. For example, if the user sees a lamp flash and asks "what is this?" the system will be able to answer "Please, stop immediately. You have run out of oil" or provide a set of suitable actions, because this event has been previously logged and an ontology describes what actions to take in such event.

8.7.2 Semantic Web Services

More in the mid term, semantic SOA is also on its way for a wide uptake in the market. The web is evolving toward a web of services at different levels, both appealing to individuals and organizations. However, these services need to be exposed to be perceived outside their respective organizations. Semantic technologies are key in order to produce comprehensive, semantic descriptions of these services providing fine-tuned service discovery and combination capabilities. The impact of SOA applications in large corporations has been demonstrated already and will continue in the future.

On the other hand, semantic technology providers have a much more realistic business opportunity in a scenario biased toward the integration of different distributed applications and services on the side of final users, rather than on the integration of the business processes of large corporations. This kind of service mash-ups will be reflected on virtual desktops that will support us both at home and work, helping us to manage our tax payment, plan our vacation, or arrange a business meeting in Singapore with the best flight connections available. Once again, as for natural interaction, customization based on context-awareness and user profiling will be a key issue.

However, there is a trust issue to be solved. How can we be sure that these services actually do behave like they are supposed to? Provenance will play a fundamental role with regard, to this increasing understanding of the interactions between services by explaining process execution in a closer way to how users perceive or reason on a given problem and monitor their execution. This goal will be supported by semantic components like problem-solving methods and ontologies that support interpretation of the execution of past processes composed by the aggregation of different services.

8.7.3   Collaborative Scenarios for Semantic Applications

The philosophy of Web 2.0 is simply orthogonal to natural interaction and SOA and intersects with both. As a "democratic" model vs. the "oligarchic" model of Web 1.0, Web 2.0 proposes a scenario where potentially everybody with access to the Internet can not only consume but also produce online content both as data or services. This is reflected in both scenarios, natural interaction and semantic SOA, where flagships of Web2.0 like customization, context-awareness, collaboration or mash-ups shine on their own.

Semantic technologies are experimenting a new boost, resulting from the basic needs of human beings to collaborate and express their ideas, which are combining with already existing applications of the technology, now stimulated by this paradigm shift and resulting in new business opportunities. However, this will bring new issues that need to be solved. For example, the appearance of micro-formats has proved how far a little semantics can go but what if the total amount of Web 2.0 heterogeneous resources, a blend of informal knowledge repositories e.g. tag clouds, folksonomies, etc, that stem from the collaboration between zillions of individuals dramatically increases, is expected? Will the current Web 2.0 devices scale appropriately? Quite probably not. The formal methods of Semantic Web is introduced in order to grant comprehensive information access and management and we will be talking about a new paradigm shift: Web 3.0.

Some of these approaches with two real-world applications will be illustrated in two different scenarios: public administrations and citizens service, and interoperability in eBusiness.

8.7.4   Semantic Applications in Public Administrations

Since some years, most Spanish local city governments offer their citizens the possibility to access and execute government services such, and from 2006, citizens have the possibility to perform 80 per cent of the city government services from their homes. However, when a citizen is looking for a particular service, it turns out that it is not so easy to find the appropriate service in the website of the City. First of all, many city websites simply enumerate the available services in a list organized by categories. In case there are 500 or 1000 services (see UK's Local Government Service List) to choose from, it may be difficult to find the right one. Other city websites offer a traditional search engine that retrieves services based on co-occurrence of words in the query and the description of the services. However, the language used by Public Administrations is not always the same as the way citizens refer to services, and there may be many ways to ask for the same thing. For instance, when a citizen wants to throw away an old washing machine, one needs to know that the government service is called "special collections for large items".

In order to stimulate the uptake of eGovernment [9, 33, 51] it is therefore very important to access online services as easy as possible. Disappointed citizens are not likely to return to their local municipality website for other services and information.

This application uses ontologies and Semantic Web technology to improve citizens' access to online services. Two important improvements concern:

- City governments should not force citizens to learn their jargon
- When citizens look for a particular service, it is useful to find also related services (e.g. if I change address, I may want to apply this change also to my car).

The city of Zaragoza is pioneer in the adoption of this initiative, being a reference for the Spanish municipalities with this regard. In fact, it has been selected as the most innovative city government for the second consecutive year (2005-06) by the Spanish renowned national newspaper El Pais [42]. More details on this work can be found in [17]. In contrast to traditional search engines that retrieve documents based on occurrences of keywords, this approach has some understanding of online services. It knows for instance that persons can change address, car owners need to pay taxes, certain business may cause hinder (such as bars and discotheques), and that there are different kinds of reconstruction works each requiring different permits.

All this information is stored in a so-called ontology: a computer-understandable description of what e-government services are. This ontology allows our product to "understand" what citizens ask for and in response to carry out the relevant services. In addition, it returns related services to the those requested. The main parts of the ontology are the following (see also Figure 8.5):

- **Agent** participating in an action
- **Process:** A series of actions that a citizen can do using the on line services offered by the city government.
- **Event:** Any social gathering or activity.
- **Object:** Any entity that exists in the city, which can be used for or by a service offered by the city government.

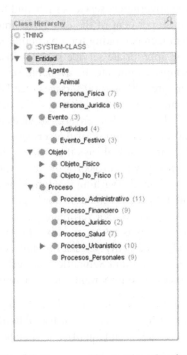

Fig. 8.5 Zaragoza City Services Ontology

In this kind of systems, the next three components are key. A domain ontology, which provides vocabulary for the description of city services, a Natural Language Processing (NLP) system, used to analyze user utterance, classifying the terms contained within as occurrences of ontology terms, and finally, a semantic distance component, which checks these terms against others, previously stored, and describe the services supported by the system. On the other hand, these components are used throughout two different stages: i) the semantic annotation of the city services about which the system needs to provide information to citizens and ii) the current semantically-enabled search of city services that passes through a set of stages (an ontology domain detection stage, keyword matching stage, and an ontology concept graph path matching stage).

Next, some examples of questions satisfied by the intelligent search engine deployed in the city of Zaragoza are illustrated below.

8.7.4.1   Some Search Examples

In this query, a user expresses interest about the city procedures related with a change of residence. Note that the language in which the question is formulated is informal, certainly far from the technical language used in official public administration documents. In this case, the relevant terms of the question with respect to the ontology and the thesauri are *mudarme* as a synonym of *mudar*, i.e. to move, and *casa* (house, in English). As a reply, the system identifies two services focused on how to obtain permission from the city to actually moving and the steps necessary to put into order and update the new postal address in the city registry, respectively.

Resultados de la Búsqueda   *quiero mudarme de casa*

> 1: Permiso para Mudanzas
> 2: Padrón Municipal de Habitantes. Cambio de domicilio

Fig. 8.6 Sample question 1: I want to change my residence

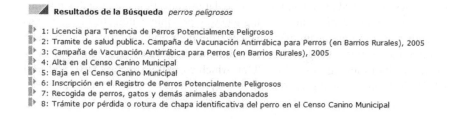

Resultados de la Búsqueda   *perros peligrosos*

> 1: Licencia para Tenencia de Perros Potencialmente Peligrosos
> 2: Tramite de salud publica. Campaña de Vacunación Antirrábica para Perros (en Barrios Rurales), 2005
> 3: Campaña de Vacunación Antirrábica para Perros (en Barrios Rurales), 2005
> 4: Alta en el Censo Canino Municipal
> 5: Baja en el Censo Canino Municipal
> 6: Inscripción en el Registro de Perros Potencialmente Peligrosos
> 7: Recogida de perros, gatos y demás animales abandonados
> 8: Trámite por pérdida o rotura de chapa identificativa del perro en el Censo Canino Municipal

Fig. 8.7 Sample question 2: dangerous dogs

This query aims at obtaining all the information about city procedures dealing with dangerous dogs. The relevant terms are *perro* (dog) and *peligroso* (dangerous). The system reply contains all the city procedures and services related with dangerous dogs. Note that the results are ordered in terms of their relevance with respect to the user query.

8.7.5   Semantic Applications in eBusiness

Since a European directive back in 2002 authorized the use of digitally signed electronic invoices for commercial transactions, the use of electronic invoicing has grown exponentially. However, its main adoption barrier has been identified as the heterogeneity of the means to represent and exchange invoice information, illustrated by the large amount of vendors of electronic invoicing technology (SAP, Oracle, PeopleSoft, Baan, Movex, etc) and languages for invoice representation (EDIFACT, UBL, XBRL, IDOC, etc), as well as the slow adoption of invoice standards

(EDI[6]) by the main players of the sector. Indeed, companies are receiving electronic invoices from many other organizations, each of which are potentially using a different format and their own electronic invoice language/technology.

The PharmaInnova[7] cluster is an initiative that groups the most important pharmaceutical laboratories in Spain, and intends to stimulate the uptake of electronic invoicing in the pharmaceutical sector by adopting a common electronic invoice model and to platform with the twofold goal of reducing costs and maximizing resilience to change in mainstream electronic invoicing technology. However, in order to adopt the solution provided by PharmaInnova, member laboratories and the organizations holding commercial transactions with them need to either substitute their existing invoicing platforms and model with PharmaInnova's or alternatively design and implement ad-hoc transformations between each pair of electronic invoice formats and models potentially participating in economic transactions. Since specialized IT staff is required with a deep knowledge of the electronic invoicing domain, both approaches are expensive and cumbersome

By using i2Ont, a NeOn-enabled application for electronic invoicing based on networked ontologies, domain experts (financial and supply chain management staff) can define themselves the necessary mappings between their own electronic invoice model and to format a conceptual, formal model provided by networked ontologies developed in NeOn. For a particular organization, creating this mapping makes the use of i2Ont as an invoicing gateway possible, automatically and transparently transforming the emitted invoices from their original format and model to those of the receiver, using networked ontologies as a backbone. Figure 8.8 shows the network of ontologies used in i2Ont, which currently supports UBL and EDIFACT electronic invoicing standards and the proprietary model of PharmaInnova.

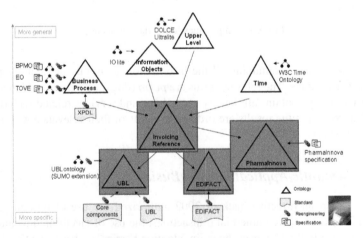

Fig. 8.8 i2Ont Ontology Network

---

[6] http://en.wikipedia.org/wiki/Electronic_Data_Interchange
[7] http://www.pharmainnova.com

One of the major challenges that needs to be tackled by i2Ont is the use by domain experts of ontology-related technologies. For this reason, a user friendly interface is developed, allowing users to navigate the invoice models and create mappings between local models and the reference one (the underlying ontology) with minimal knowledge of the reference model and concealing the complexity of ontology technologies as much as possible (see Figure 8.9). More details on this application and on how NeOn technology has been used in its development can be found in [18].

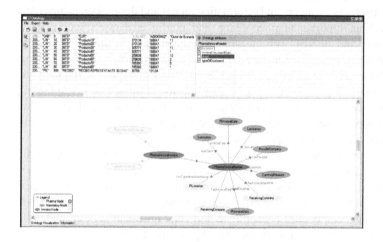

Fig. 8.9 The i2Ont Electronic Invoicing Application

8.8   New Challenges: From Semantic Islands to the Semantic Cloud

In previous sections, the state of the at in Ontological Engineering and Semantic Web have been described from different perspectives ranging from knowledge representation and reasoning frameworks to networked ontologies. Furthermore, such approaches have been illustrated with two semantic web applications using the previously described breakthroughs. However, although these new approaches are certainly promising and reaching the market, nowadays we are facing a world of opportunities and challenges, as mentioned in the introduction. In this section, we will briefly explain some of these opportunities and challenges.

In the early days of the Semantic Web, the task of producing semantic content required large amounts of effort in order to develop tools and methods capable of providing existing web content with semantics. Nevertheless, applications like semantic search engines and semantically-enabled mash-ups of services were produced, as seen in section 8.7. Companies indeed have successfully started to deploy their products, fundamentally in the corporate environment and in other sectors such as finance and public administration.

However, the situation has now reached an inflexion point. Semantic resources are no longer isolated silos but, on the contrary, the abundance of such resources is facilitating the appearance of initiatives like the Linked Open Data[8] whose main goal is to connect these information islands in order to accomplish the Semantic Web as originally envisioned. The sheer amount of semantic descriptions and the different perspectives addressed by such descriptions will stimulate a new generation of intelligent, multifaceted applications with richer content and services. As pointed out by Leigh Doods[9], using the Linked Data principles for intelligent management of information one can form a mesh of interrelated facts and figures that can help others to understand information pieces and their relationships with others.

While an important number of large and interlinked datasets, such as the UK government[10] and the BBC web site, are beginning to gain worldwide popularity and are currently available by addresing the Linked Data principles, several challenges need to be addressed before this can be achieved. New techniques are required which, beyond controlling and securing access and edition of the semantic resources, provide the mechanisms needed to sustain their trust, quality, and timeliness. Provenance-aware applications are therefore needed to help identify trustworthy content and services, allowing their appropriate (re)use.

First, new technologies are required to support the (semi)automatic identification, extraction, and interconnection of relevant information from amidst large volumes of online information. The resulting information networks need to be context-aware and user-oriented, providing information consumers with contextualized views and navigation metaphors and adapting to particular information needs and preferences. Users are enabled to find information objects that draw on the same set of assumptions and to discover related information and alternative viewpoints, eventually creating a rich information fabric. Likewise, organizations revealing their literature online need to be provided with the means that allow them to enrich their information assets, dynamically linking them to such fabric, benefiting from previously existing and potentially complementary contents, and increasing throughput in the generation of modern information.

Second, linking to external information without the appropriate means that allow contrasting its provenance can be harmful, especially in sensitive domains. For example, in September 2009 Germany was shocked by a scheme that managed to have the Deutsche Presse Agentur[11] (DPA) report on the attempted suicide attack supposedly perpetrated by a non-existing terrorist group called Berlin Boys[12] in a city called Bluewater, California. In fact, this turned to be a hoax organized by a group of German filmmakers trying to win the attention of the media to promote their work.

The elaborated hoax involved at least two fake websites, a fake Wikipedia entry and a list of California phone numbers of public safety officials that were being

[8] http://linkeddata.org
[9] http://blogs.talis.com/nodalities/2009/07/linked-data-and-news-innovation.php
[10] http://news.bbc.co.uk/2/hi/technology/8096793.stm
[11] http://www.dpa.de
[12] http://www.wired.com/threatlevel/2009/09/bluewater

contacted by hoaxers in Germany through Skype. Eventually, the hoax prompted a 1000-word tome on the Frankfurter Allgemeine website and forced DPA to publicly apologize for the incident. In a provenance-aware world, DPA would have had the means to automatically reason on the provenance of the information and therefore would have realized that the city itself did not in fact exist despite appearing in Wikipedia, used by DPA as a trustworthy information source.

Information reliability cannot be built exclusively from arbitrary notions of trust. On the contrary, trust needs to be built through the appropriate analysis of the provenance of information, supporting decision making in the light of establishing its soundness. According to Tim Berners-Lee[13]. Building systems that track its sources and the way in which information was obtained by them is more important and practical than trying to control and secure the information flow. In the Linked Data paradigm, provenance shall provide the ability to trace the sources of data, enabling the exploration not just of the relationships existing between information objects, but also of their authors and affiliations. Through the analysis of the provenance of information, context is gained not just regarding the information itself, but also on the way in which (How) it is produced and exploited.

Acknowledgements. This work has been funded by IST-2005-027595 EU project NeOn . Our gratitude goes to the members of the NeOn consortium. To a large extent, this chapter summarizes part of the team work developed in the project during the last four years.

References

1. Auer, S., Dietzold, S., Riechert, T.: OntoWiki-A Tool for Social, Semantic Collaboration. In: Cruz, I., Decker, S., Allemang, D., Preist, C., Schwabe, D., Mika, P., Uschold, M., Aroyo, L.M. (eds.) ISWC 2006. LNCS, vol. 4273, pp. 736–749. Springer, Heidelberg (2006)
2. Baader, F., Sattler, U.: An overview of tableau algorithms for description logics. Studia Logica 69, 2001 (2000)
3. Bao, J., Caragea, D., Honavar, V.: Towards collaborative environments for ontology construction and sharing. In: International Symposium on Collaborative Technologies and Systems (CTS 2006), pp. 99–108 (2006)
4. Bechhofer, S., van Harmelen, F., Hendler, J., Horrocks, I., McGuinness, D.L., Patel-Schneider, P.F., Stein, L.A.: OWL Web Ontology Language
5. Berners-Lee, T., Hendler, J., Lassila, O.: The Semantic Web. Scientific American (2001)
6. Borgida, A., Serafini, L.: Distributed description logics: Directed domain correspondences in federated information sources. In: Meersman, R., Tari, Z., et al. (eds.) CoopIS 2002, DOA 2002, and ODBASE 2002. LNCS, vol. 2519, pp. 36–53. Springer, Heidelberg (2002)
7. Borst, W.N.: Construction of Engineering Ontologies. Centre for Telematic and Information Technology, University of Tweenty, Enschede, The Netherlands (1997)
8. Bouquet, P., Giunchiglia, F., van Harmelen, F., Serafini, L., Stuckenschmidt, H.: COWL: Contextualizing ontologies. In: Fensel, D., Sycara, K., Mylopoulos, J. (eds.) ISWC 2003. LNCS, vol. 2870, pp. 164–179. Springer, Heidelberg (2003)

---

[13] http://www.w3.org/2007/09/21-timbl-egov

9. European Commission. eEurope 2005 Action Plan (2005)
10. d'Aquin, M., Schlicht, A., Stuckenschmidt, H., Sabou, M.: Ontology modularization for knowledge selection: Experiments and evaluations. In: Wagner, R., Revell, N., Pernul, G. (eds.) DEXA 2007. LNCS, vol. 4653, pp. 874–883. Springer, Heidelberg (2007)
11. David, J., Guillet, F., Briand, H.: Association rule ontology matching approach. International Journal on Semantic Web and Information Systems 3(2), 27–49 (2007)
12. Dellschaft, K., Engelbrecht, H., Barreto, J.M., Rutenbeck, S., Staab, S.: The Semantic Web: Research and Applications. In: Cicero: Tracking Design Rationale in Collaborative Ontology Engineering, pp. 782–786. Idea Publishing Group, USA (2008)
13. Domingue, J.: Tadzebao and WebOnto: Discussing, Browsing, and Editing Ontologies on the Web. In: Proceedings of the 11th Knowledge Acquisition for Knowledge-Based Systems Workshop (1998)
14. Doran, P., Tamma, V., Iannone, L.: Ontology module extraction for ontology reuse: An ontology engineering perspective. In: Proceedings of the 2007 ACM CIKM International Conference on Information and Knowledge Management (2007)
15. Farquhar, A., Fikes, R., Rice, J.: Ontolingua Server: a tool for collaborative ontology construction. International Journal of Human-Computers Studies 46(6), 707–727 (1997)
16. Fernández-López, M., Gómez-Pérez, A., Juristo, N.: METHONTOLOGY: From Ontological Art Towards Ontological Engineering. In: Spring Symposium on Ontological Engineering of AAAI, pp. 33–40. Stanford University, Stanford (1997)
17. Gómez-Pérez, J.M., Benjamins, V.R., Blázquez, M., Contreras, J., Fernández, M.J., Patón, D., Rodrigo, L.: An Intelligent Search Engine for Online Access to Municipal Services. In: ESTC 2007 (2007)
18. Gómez-Pérez, J.M., d'Aquin, M., Haase, P.: NeOn - Lifecycle support for Networked Ontologies A Case Study in the Pharmaceutical Domain. In: ESTC 2008 (2009)
19. Cuenca Grau, B., Horrocks, I., Kazakov, Y., Sattler, U.: A logical framework for modularity of ontologies. In: Proc. of 20th International Joint Conference on Artificial Intelligence (IJCAI 2007), pp. 298–303 (2007)
20. Cuenca Grau, B., Kutz, O.: Modular ontology languages revisited. In: Workshop on Semantic Web for Collaborative Knowledge Acquisition, SWeCKa 2007 (2007)
21. Cuenca Grau, B., Parsia, B., Sirin, E.: Working with multiple ontologies on the semantic web. In: McIlraith, S.A., Plexousakis, D., van Harmelen, F. (eds.) ISWC 2004. LNCS, vol. 3298, pp. 620–634. Springer, Heidelberg (2004)
22. Gruber, T.: A translation approach to portable ontology specifications. Knowledge Acquisition 5, 199–220 (1993)
23. Guarino, N.: Formal Ontology in Information Systems. In: 1st International Conference on Formal Ontology in Information Systems (FOIS 1998), pp. 3–15 (1998)
24. Haase, P., Rudolph, S., Wang, Y., Brockmans, S., Palma, R., Euzenat, J., d'Aquin, M.: Networked Ontology Model v1.0
25. Hitzler, P., Krötzsch, M., Parsia, B., Patel-Schneider, P.F., Rudolph, S.: OWL2 Web Ontology Language
26. Jean-Mary, Y., Kabuka, M.: ASMOV: Ontology Alignment with Semantic Validation. In: Joint SWDB-ODBIS Workshop, pp. 15–20 (2007)
27. Kahan, J., Koivunen, M.R., Prud Hommeaux, E., Swick, R.R.: Annotea: an open RDF infrastructure for shared Web annotations. Computer Networks, 589–608 (2002)
28. Kalyanpur, A., Parsia, B., Sirin, E., Grau, B.C., Hendler, J.: SWOOP: A Web Ontology Editing Browser. In: Web Semantics: Science, Services and Agents on the World Wide Web, pp. 144–153 (2006)
29. Kaushik, S., Farkas, C., Wijesekera, D., Ammann, P.: An algebra for composing ontologies. In: Formal Ontology in Information Systems, FOIS 2006 (2006)

30. Kutz, O., Lutz, C., Wolter, F., Zakharyaschev, M.: E-connections of description logics. In: Description Logics Workshop, CEUR-WS, vol. 81 (2003)
31. Lambrix, P., Tan, H.: SAMBO - a system for aligning and merging biomedical ontologies. Journal of Web Semantics, Special issue on Semantic Web for the Life Sciences 4(3), 196–206 (2006)
32. Li, K., Xu, B., Wang, P.: An Ontology Mapping Approach Using Web Search Engine. Journal of Southeast University 23(3), 352–356 (2007)
33. Liikanen, E.: e-Gov and the European Union (2003)
34. de Cea, G.A., Suárez-Figueroa, M.C., Buil, C., Caracciolo, C., Dzbor, M., Gómez-Pérez, A., Herrrero, G., Lewen, H., Montiel-Ponsoda, E., Presutti, V.: Deliverable D5.3.1 NeOn Development Process and Ontology Life Cycle (August 2007)
35. MacCartney, B., McIlraith, S., Amir, E., Uribe, T.E.: Practical Partition-Based Theorem Proving for Large Knowledge Bases. In: Proc. of the International Joint Conference on Artificial Intelligence, IJCAI 2003 (2003)
36. Melnik, S., Bernstein, P.A., Halevy, A.Y., Rahm, E.: A semantics for model management operators (2004)
37. Melnik, S., Rahm, E., Bernstein, P.A.: Rondo: A programming platform for generic model management. In: SIGMOD, pp. 193–204 (2003)
38. Nagy, M., Vargas-Vera, M., Motta, E.: DSSim - Managing uncertainty on the Semantic Web. In: Proceedings of the 2nd International Workshop on Ontology Matching (2007)
39. Neches, R., Fikes, R., Finin, T., Gruber, T., Senator, T., Swartout, W.: Enabling technology for knowledge sharing. AI Magazine 12(3), 36–56 (1991)
40. Newell, A.: The Knowledge Level. Artificial Intelligence 18(1), 87–127 (1982)
41. Noy, N.F., Musen, M.A.: Specifying Ontology Views by Traversal. In: McIlraith, S.A., Plexousakis, D., van Harmelen, F. (eds.) ISWC 2004. LNCS, vol. 3298, pp. 713–725. Springer, Heidelberg (2004)
42. El País. Todos los grandes ayuntamientos tienen web
43. Pan, J.Z., Serafini, L., Zhao, Y.: Semantic import: An approach for partial ontology reuse. In: Workshop on Modular Ontologies (2006)
44. Pinto, H.S., Tempich, C., Staab, S.: DILIGENT: Towards a fine-grained methodology for DIstributed, Loosely-controlled and evolvInG Engineering of oNTologies. In: Proceedings of the 16th European Conference on Artificial Intelligence (ECAI 2004), pp. 393–397 (2004)
45. Seidenberg, J., Rector, A.: Web ontology segmentation: Analysis, classification and use. In: Proceedings of the World Wide Web Conference, WWW 2006 (2006)
46. Staab, S., Schnurr, H.P., Studer, R., Sure, Y.: Knowledge Processes and Ontologies. IEEE Intelligent Systems 16(1), 26–34 (2001)
47. Stuckenschmidt, H., Klein, M.C.A.: Structure-based partitioning of large concept hierarchies. In: McIlraith, S.A., Plexousakis, D., van Harmelen, F. (eds.) ISWC 2004. LNCS, vol. 3298, pp. 289–303. Springer, Heidelberg (2004)
48. Studer, R., Benjamins, R., Fensel, D.: Knowledge Engineering: Principles and Methods. IEEE Transactions on Data and Knowledge Engineering 25(1-2), 161–197 (1998)
49. Swartout, B., Patil, R., Knight, K., Russ, T.: Ontosaurus: a tool for browsing and editing ontologies. In: 9th Knowledge Aquisition for Knowledge-based systems Workshop (1996)
50. Tang, J., Li, J., Liang, B., Huang, X., Li, Y., Wang, K.: Using Bayesian Decision for Ontology Alignment. Journal of Web Semantics 4(4), 243–262 (2006)
51. Tarabanis, K., Peristeras, V.: Requirements for Transparent Public Services Provision amongst Public Administrations. In: Traunmüller, R., Lenk, K. (eds.) EGOV 2002. LNCS, vol. 2456, pp. 330–337. Springer, Heidelberg (2002)

52. Tudorache, T., Noy, N.: Collaborative Protégé. In: Workshop on Social and Collaborative Construction of Structured Knowledge colocated with WWW 2007 (2007)
53. Tummarello, G., Morbidoni, C., Nucci, M.: Enabling Semantic Web communities with DBin: an overview. In: Cruz, I., Decker, S., Allemang, D., Preist, C., Schwabe, D., Mika, P., Uschold, M., Aroyo, L.M. (eds.) ISWC 2006. LNCS, vol. 4273, pp. 943–950. Springer, Heidelberg (2006)
54. Volz, R., Oberle, D., Maedche, A.: Towards a Modularized Semantic Web. In: Semantic Web Workshop (2002)
55. Wiederhold, G.: An algebra for ontology composition. In: Monterey Workshop on Formal Methods (1994)
56. Zimmermann, A.: Integrated Distributed Description Logics. In: Proc. of the 20th International Workshop on Description Logics, DL 2007 (2007)

Chapter 9
Web Intelligence on the Social Web

Sebastián A. Ríos and Felipe Aguilera

Abstract. The WWW, has become a fertile land where anyone can transform his ideas into real applications to create new amazing services. Therefore, it was just a matter of time until the massive proliferation of virtual communities, social networks, etc. New social structures have been formed by massive use of new technologies. This way, people can relate to other by interests, experiences or needs. In a scenario where WWW has become more important every day, and people is using more often the web to relate to others, to read news, obtain tickets, etc. The need of well organized web sites has become one of the vital goals of enterprises and organizations. To accomplish such task web mining area was born more than a decade ago. Web mining are techniques that help managers (or web sites' experts) to extract information from a web sites' content, link structure or visitors' browsing behavior. This way, it is possible to enhance a web site, obtain visitors' interests patterns to create new services, or provide very specific adds depending on the navigation preferences of visitors (recommendations systems). In the beginning of the Web, web sites were formed by static pages, this means contents were created usually by the owner of the web sites, or the web masters. These contents usually did not change very much through time since it required effort from administrators. Today, a new paradigm arose, we have a participative Web. The web has evolved to the point that it is composed by dynamic contents created by millions of users collaborating one to each other. Sites like, youtube, Blogger, Twitter, facebook, orkut, flickr, among many other, are part of the social web sites' phenomenon. For example, twitter had $475,000$ members by Feb. 2008 while it had $7,038,000$ members by Feb. 2009, which means 1382% of growth. Facebook on the same dates passed

Sebastián A. Ríos
Department of Industrial Engineering, University of Chile, República 701, Santiago, Chile
e-mail: srios@dii.uchile.cl

Felipe Aguilera
Department of Computer Science, University of Chile, Av. Blanco Encalada 2120, Santiago, Chile
e-mail: faguiler@dcc.uchile.cl

J.D. Velásquez and L.C. Jain (Eds.): Advanced Techniques in Web Intelligence – 1, SCI 311, pp. 225–249.
springerlink.com   © Springer-Verlag Berlin Heidelberg 2010

from 20,043,000 members to 65,704,000 members which means 228%. The use of web intelligence techniques to explode data stored in these social web has become a natural approach to obtain knowledge from them. Since volumes of data are huge, the use of web intelligence techniques was the natural approach to obtain knowledge from social web sites. However, to study members of a social web site is not only to study a group of people accessing a web site and working together; they establish social relationships through the use of Internet tools allowing the formation shared identity and a shared sense of the world. In order to provide truly valuable information to help managers, web masters and to provide better members' experience when using the social web site, it is necessary to take into account datas' social nature in web mining techniques. This chapter focuses on the application web intelligence techniques in combination to social network analysis to study of social web sites. In order to provide truly valuable informaton from social web sites that support a social entity. We show that new techniques need to be focused on the study of underlying social aspects of those social entities to really exploit the datas' social nature and provide a better understandig of human relationships.

9.1 Introduction

In a scenario where WWW has become more important every day, and people is using more often the web to relate to others, to read news, obtain tickets, etc. The need to well organized web site has become one of the vital goals of enterprises and organizations. In order to accomplish such task web mining area was born about a decade ago. The WWW has also become a fertile land where anyone can transform his ideas into real applications to create new amazing services. Therefore, it was just a matter of time until the massive proliferation of virtual communities, social networks, community tagging systems, wikis, forums, etc. New social structures have been formed by massive use of new technologies. This way, people can relate to other by interests, experiences or needs.

Web mining are techniques to help managers (or experts) to extract information form a web site content, link structure or visitors' browsing behavior. This way, it is possible to enhance a web site, obtain visitors' interests patterns to create new services, or provide very specific adds depending on interests of visitors. Web mining, is based on the Knowledge Discovery in Databases (KDD) process and uses data mining techniques to extract information, discover patterns, etc. Literature on web mining abound and some good references could be found in [10, 34, 43, 46, 48, 47, 68, 73, 76].

Today, virtual communities have experienced an exponential growth. Also, the use of web mining techniques to explode data stored in these systems has become a natural approach to obtain knowledge from them [24, 73, 28, 44, 49, 75]. However, a virtual community is not only a group of people accessing a web site, they establish social relationship through the use of Internet tools [83], allowing the formation of a communal identity and a shared sense of the world [84]. In order to provide truly valuable information to help managers or web masters it is necessary take into account the social nature of the virtual communities in web mining techniques.

The aim of this chapter is to show that a social web site is used by a social structure (like a virtual community). Therefore, the application of web mining techniques to study this social structures is not enough. To obtain a complete description of the inner social structure behind the social web site, it is needed the use of techniques to study social aspects of this social structure. We focused the chapter mainly in characterize the social entities by it social aspects. Then we explain state-of-the-art on techniques to study social entities. Then we show how web mining techniques is used to study social web sites in a traditional way. Afterwards, we show how social techniques can be improved by the combination with web mining techniques.

9.2   Social Aspects on Communities and Social Networks

Massive use of Internet services such as web systems, allowed that many people could communicate and interact with other, without concerning its geographic location. It is possible to converse with other people, find people with common interests to help others in certain problems, share information, participate in discussions, etc.. These activities have led to the use of the computer goes from being an individual activity to a collective activity, which create different links of interaction and cooperation with others.

The importance of Internet, has led to the emergence of new social institutions [83, 80] which, although, are based on existing ones, posseses specific characteristics [31] which need to be considered when performing its study.These differences are due to the use of a different medium to the interaction face-to-face, which generates many social rituals that exist in the real world [82] does not exist or are limited in the virtual world [29].

This way, have surfaced new forms to relate between people, around these new social structures: social networks, virtual communities, virtual communities of practice, etc. Understand what are the main aspects of these social structures is essential to support them properly.

In this section will explain the most important social structures and we developed summary of their social aspects in Table 9.1.

9.2.1   Online Social Networks

Internet has had the strength to not only connect computers across the network, but also allowed connecting people [8, 78]. The use of email, discussion forums, and other systems has enabled people to stop working in isolation, and beging to work in online groups, facilitated by the formation of social networks.

A network is basically a set of objects (called nodes) and a set of relationships between these objects. Based on the foregoing, a social network is a set of social institutions associated with any social character, such as friendship, co-working or

Table 9.1 Social Aspects in Social Structures (Social networks, Virtual Communities, Virtual Communites of Practice, among other)

Author	Social Aspect	Description
Social Networks		
Wellman [23]	Range	It refers to their size and heterogeneity.
	Centrality	it refers to who is central or isolated in network.
	Roles	It refers to similarities in network members' behavior.
	Groups	Is an empirically discovered structure. By examining the pattern of relationships among members of a population, groups emerge as highly interconnected sets of actors known as cliques and clusters.
Virtual Community		
Preece [53]	People	The people who interact with each other in the community and who have individual, social and organization needs.
	Purpose	A communitys shared focus on an interest, need, information, service, or support, that provides a reason for individual members to belong to the community.
	Policies	The language and protocols that guide people's interactions and contribute to the development of folklore and rituals that bring a sense of history and accepted social norms.
Kim [31]	Purpose	This is when we focus on the sheared goals of community members
	Places	It refers to infrastructure of gathering places, which members work together to evolve.
	Profiles	It is a collection of information that says something about who a member is in the context of the community
	Roles	These are the different types of members in a community.
	Leadership	It refers to most visible role models. Volunteers, contractors, and staff that keep the community running
	Etiquette	It is a set of behaviors -or community standards- that a group of people has agreed to abide by.
	Events	Every long-lasting community is brought together by regular events. These gatherings help define the community, and they remind people of what they have in common and what their community is all about.
	Rituals	These are social acts' celebrations (v.g. a members' birthday), which allow developement of a true on-line culture (identity and belongness of/to a group).
	Subgroups	It refers to small groups where people form their deepest relationships and strongest loyalties.
Community of Practice		
Wenger [87]	Community	It refers to builded relationships that enable members to learn from each other
	Practice	It refers to a shared repertoire of resources: experiences, stories, tools, ways of addressing recurring problems
	Domain	A CoP has an identity defined by a shared domain of interest.
Plaskoff [52]	Believing	It refers to the idea that members need to believe in the intrinsic value of a community.
	Behaving	It indicates that members develop and follow norms of a community.
	Belonging	It means that members nurture a sense of belonging within a community.

information exchange. These entities could be from individuals to groups, or other organization.

A group of people who interact online, implicitly build a social network, though not be defined explicitly [8].

Under the social network perspective, the following aspects are crucial:

- The actors and their actions are seen as interdependent, rather than seeing them as autonomous and independent units.
- Relations between actors are channels that allow resources transference/flow (material or not).

9.2.2   Virtual Communities

Find a precise definition of community is not a simple task. One may find many definitions, from a social perspective – which have being evolving and redefined over time [3, 39]. However, there are indicators of community that have been adopted by many researchers: the concept of people sharing some common interests, experiences and / or needs, linked through social relations, through which important resources are found, they develop strong interpersonal feelings of belonging and mutual need, anda sense of shared identity emerge [30, 57, 79].

The types of communities created through Internet differed in several respects from the traditional concept of community, mainly due to the fact that they are mediated through computers. These types of communities are called virtual communities or online communities. In a virtual community people can meet face to face with others or not, and the exchange of messages and ideas takes place through the network [57].

Virtual communities can be seen as a social network, since the network of computers that connects people is able to create a set of links with social meaning [78] among people . Similarly, a working group is also considered a social network, the main difference with a virtual community is that the relations between people are strongly bounded on its border (there are no relations with people outside the working group) and they are very dense (each person is related to the majority of the members of the group) [78]. A virtual community is created from the continuity of relations between its members, but it is experienced through specific activities localized in time and space [89].

The main types of communities studied today are as follows:

- *Communities of interest [38].* Are those communities in which members share the same interest in any topic (and therefore they all have a common background.) Examples of this type of community: music bands fan club, groups of people interested in the planets of the solar system, groups of people interested on environment, among many other. For some authors communities of passion [9] exhibit a very similar definition to interests communities.
- *Communities of practice [70, 85, 88].* Are those communities in which members share the same profession or activity. Usually its members are heavily involved in the community. Examples of this type of community are the Smalltalk

development community, The Open Source community, communities inside enterprises are generally of these kind, etc.

- *Communities of purpose [9]*. Are those communities in which its members share the same objective (usually short term). Examples of this type of community are the buyers of a virtual library, which share the goal of finding and buying a book. Members have a functional purpose which is disbanded once the goal is reached. Usually members of this type of community does not engage in activities that exceed the purposes of the community and nor share the same interest [9].

There are other types of communities, however, we won't go further in this chapter. An interested reader may review the following references [3, 25].

9.2.3 Virtual Communities of Practice

Virtual Communities of Practice (VCoP) have experienced an explosive growth in Internet in recent years. VCoPs' value is to communicate people who wants to share or learn about a specific topic, by interacting on an ongoing basis [87]. The Web facilitates interaction between community members without presentational contact needed, ubiquitous environment and atemporal virtual communication space. To support this virtual interaction are commonly used forums, wikis, and other similar tools.

For the VCoPs is very important to generate, store and keep knowledge resulting from members' interaction. The success of a VCoP depends on a governance mechanism [56] and key members' participation (so called leader [5] or core members [56]). Likewise, every VCoP members' goal is to learn specific knowledge from the community, therefore, the contents of posts, which specific topic is interesting for a member, etc. must be considered.

Plaskoff [52] has suggested a means for fostering communities of practice based on his own experience in a large pharmaceutical company. He notes three concepts that must exist for the successful cultivation of a community of practice within an organization: believing, behaving, and belonging. Believing refers to the idea that members need to believe in the intrinsic value of a community. Behaving indicates that members develop and follow norms of a community. Belonging means that members nurture a sense of belonging within a community.

There are different kinds of communities. Kim et al. [32] organize Social Web Communities describing the kind of users, uses and needed features for every kind of community. Important missions for a Community of Knowledge are sharing user-created content and how to hold users, specially key members. In this context, VCoP is also a Community of Knowledge, therefore, it should accomplish such missions, but also, it must care that community members fulfil their goals (or purpose) when using the VCoP.

9.3   Social Networks and Virtual Communities Analysis Techniques

9.3.1   Social Network Analysis

The area of Social Network Analysis aims to study social structures modeled in a network form, i.e. as a set of actors and social relations between these actors. Unlike other types of analysis, SNA is not linear [22], that is, their analysis can not be done incrementally, but must be done in the complete structure [67]. These properties are due to SNA are based on a structural analysis [81], i.e. the analysis is based on relationships between actors and interesting patterns within the network.

To perform SNA it is possible to identify six different techniques [7]:

- *Meassures*: The objective of using metrics is to measure properties owned by the social network, its actors or set of them. For example, it is possible to measure the density of a social network to determine how its members are related or is it possible to calculate the centrality of an actor, to determine their importance within the social network. Among the metrics, we can find [22, 77]:

  * *Centrality*: The degree to which an actor is in a central role in the network
  * *Prestige*: Also known as status, this measure attempt to quantify the rank that a particular actor has within a set of actors.
  * *Homophily*: The degree to which similar actors in similar roles share information
  * *Isolate*: An actor with no ties to other actors
  * *Gatekeeper*: An actor who connects the network to outside influences
  * *Cutpoint*: An actor whose removal results in unconnected paths in the network

  Among Netwroks' metrics we may find [22]:

  * *Centralization*: The fraction of main actors within a network
  * *Reachability*: The number of ties connecting actors
  * *Connectedness*: The ability of actors to reach one to another reciprocally, that is, the ability to choose a relationship between both parties
  * *Asymmetry*: The ratio of reciprocal relationships, those relationships that are mutual over total relationships within a network
  * *Balance*: The extent to which ties in the network are direct and reciprocated

- *Cohesive Subgroups*: It refers to subsets of actors among whom there are relatively strong, direct, intense, frequent, or positive ties.

  * *Mutuality Complete*: This method is based on the search for cohesive subgroups in which all social relations are possible between all actors. For them, is based on the definition of a clique, ie a maximal complete subgraph of at least three nodes.
  * *Reachibility and Diameter*: This method is based on an extension of the above method. These groups are important if we know that the most important social processes occur through intermediaries. It is based on the fact that not all

subgroup members are connected directly, however, the routes that connect them are relatively short. They are based on the definition of n-clique (is a maximal subgraph in which the largest geodesic distance between any two nodes is no greater than n; a geodesic path includes any nodes in the graph), n-clans (is a n-clique in which the geodesic distance between all nodes in the subgraph is no greater than n for paths within the subgraph) and n-clubs (is a maximal subgraph of diameter n, i.e., the distance between all nodes within the subgraph is less than or equal to n).

* *Nodal Degree*: this approach is based on restrictions on the minimum number of actors adjacent to each actor in a subgroup. It is based on k-plexes (is a maximal subgraph which each node in the subgraph may be lacking ties to no more than k subgraph members) and k-cores (is a subgraph which each node is adjacent to at least a minimum number, k, of the other nodes in the subgraph).

* *Outside Subgroup Ties*: This method is based in the idea that cohesive subgroup should be relatively cohesive within compared to outside. It is based on the definition of LS Sets (subgroup definition that compares ties within the subgroup to the ties outside the subgroups by focusing on the greater frecuency of ties among subgroups members compared to the ties from subgroups members to outsiders) and Lambda Sets (this idea extends the previous notion motivated by the idea that a cohesive subset should be relatively robust in terms of its connectivity).

* *Permutation Matrix*: The purpose of matrix permutation is to rearrange rows and columns of a matrix so that members who occupy adjacent rows (or columns) can be organized into the same group. To use the matrix permutation method a network must be represented as an $n \times n$ matrix in which both rows and columns represent n network members [90].

• *Visualization*: A social network can be modeled as a graph, representing the social actors by circles and relationships by arrows connecting the circles. This visual representation is called sociogram. These allow to explore and understand a social network, through the visual analysis of their relationships. The complexity of a sociogram lies in the technique you use to draw the social network [27].

* *Circular Layout*: All nodes are put in circles. This layout is intended to make the relationship patterns among actors salient [69].

* *Radial Layout*: Nodes are laid inside of concentric circles whose radious increase while the status levels of nodes decrease, with the most central nodes in the center of the diagram.

* *Hierarchical Layout*: maps actors' status scores to the nodes' vertical coordinates; the horizontal positioning of nodes is algorithmically computed where the overall readability in the final visual network representation is achieved.

* *Group Layout*: It is used to display information about groupings. It highlights the group existence by separating different groups and placing nodes in the same group close to each other.

* *Free Layout*: This layout does not have any purpose of highlighting a particular network feature. Instead, sociograms are drawn based on general aesthetic principles so that the layout is readable.

• *Other*

* *Role Interlock*: This technique exploite the interpenetration of multiple networks (such as liking and disliking) on a given population of actors.
* *Equivalence*: The objective is to identify sets of actors within the social network that are located in a similar way to another set of actors. Structural equivalence captures sets of actors each of whom has identical relations to all actors in a network of multiple types of ties.
* *Duality*: In [6] it is suggested the basic idea of duality of people and groups, indicating that social networks not only relate to people, but also to groups where they belong . More generally, it is possible to see social networks as networks of networks [13]. This way, people who belong to multiple groups act as bridging ties.
* *Social Influence*: The objective is to link networks of social relations to attitudes and behaviors of the actors.

9.3.2   Virtual Communities Analysis

Based on [55] we can identify four ways to analyze a VCoP which are commonly used. However, most of these approaches don't consider evolutionary factors in the analysis.

• *Ethnography and associated techniques:* The purpose of ethnographic research is to build a rich understanding of a group or situation from the point of view of its members/participants. It is a qualitative research method for understanding how technology is used *in situ* [21, 55]. An example of this technique is the application of an ethnography analysis to a virtual community that uses an IRC as communication intermediary between its members [45].

   Disadvantage of this kind of analysis is that *in situ* studies are intrusive. They are difficult to apply to communities where there are virtual members and their participation is asynchronous. Due to these problems netnography arose as a way to avoid those situations. Netnography [35] is ethnography for the study of virtual communities, commonly used in marketing field.

• *Questionnaires:* Questionnaires are useful for collecting demographic information and have the advantage that they can be distributed by hand to local participants, or posted via email or on the Web [54, 55].

   Although, questionnaires provides useful information from virtual community members [33], they are not sufficient. It is recommended the use of a secondary technique, when possible, to reduce the subjectivity of members' judgements.

• *Experiments and quasi-experiments:* Laboratory studies are valuable for testing the usability of the interface and users' reactions to new user interface features.

Such approaches could conceivably be used to investigate the impact of change in software design on online communities [54]. An example can be found in [74].

To apply this approach it is needed to create the virtual community in a laboratory. This is quite complex since it is needed to have a representative group of virtual communities members, in a real (physical) experiment. For example, some members have connection habits, which must be reproduced in laboratory. Therefore, when virtual community members use an asynchronous system or when they are remotely distributed, this approach is even harder to use.

- *Data Mining & Social Network Analysis:* This kind of analysis consist of using softwares' logs to discover useful information from the VCoP. This way, the main point is the study of communities' nature, i.e. the study of its members and the relations they establish [54].

In these studies typically two different strategies are applied. Firstly, those based on the visualization of underlying social network. As result, deduction and establishment of some behavior patterns is possible or other social structures inside the social network [2]. Secondly, those that attempt to quantify in any possible manner the relationships that are established in the community [16]. To achieve this it is a common practice to define and measure a set of measures and indicators to monitor community's behavior. These both approaches are commonly called Social Network Analysis (SNA).

9.3.3 Analysis Techniques Drawbacks

The problem with SNA, is that using any of those approaches, we end in a graph representation of the community (social network). With nodes - which represent people - and edges, which represent interactions or relations between them. However, SNA analysis is designed to answer questions like: Who is the expert or experts in the community? Which sub communities or subgroups exists? But, it is not possible to answer: which is the interests of members? which is the real goal of the community? Which topics are related with the community purpose? Which themes diverge from communities' main purpose? All these question remain unanswered. However, the answer allow managers or community owners to perform the necessary enhancements to contents and organization of VCoP, to survive through time.

We propose that SNA techniques are not sufficient to fully understand a virtual community of practice. Therefore, in this paper we propose a novel method to analyze a VCoP, based on a data mining approach to discover useful information from the VCoP for its better administration.

9.4 Web Mining on Social Web Sites

A social Web sites are Web sites that make it possible for people to form online communities, and share user-created contents (UCCs) [32]. Social web sites like Facebook, Orkut, LinkedIt, Bebo, and MySpace offer social networks' functionalities [8].

On the other hand, when data mining techniques arose they were applied to any on-line information. However, web sites at that time were static pages. The web sites evolved and social web sites arose (also called "Web 2.0"), but the need for quality information remained. Thus, web mining techniques were also applied but these techniques are not sufficient to study the social aspects (see Table 9.1) of a social web site.

Web Mining techniques arose as tools for helping managers and web masters finding interesting patterns to enhance a web site[1]. This can be done in two ways: on-line enhancements and off-line enhancements. On-line web site enhancements are changes performed in real time on the web site contents and structure [58]. The most popular example of on-line enhancements is the web site recommendations of Amazon.com. When you search for a book, Amazon.com provides the user with the recommendation of books bought by other customers. On the other hand, off-line enhancements are those applied over a copy of a web site, then an expert must understand and create changes on the contents and structure of the site. Once changes are made, this modified web site should be placed on-line. Of course, off-line and on-line changes are complementary.

Many tools provide not very useful results to perform off-line web site enhancements. A big problem is how people in charge (administrators or managers) of a web site modifies the site using web mining tools' results. The issue of analyzing huge amount of subjective information, hundreds of features, and millions of user sessions; to extract useful information to enhance a web site is not straightforward [58].

Web mining literature commonly define three main sub areas which are web text mining (WTM), web structure mining (WSM) and web usage mining (WUM). Web text mining techniques are those techniques where data mining algorithms are used to perform text classification or text clustering to find keywords (for example, WTM is used for web pages indexation), many other examples of WTM can be found in [17, 46, 48, 61, 62, 63, 65, 76]. Web structure mining tries to find interesting patterns from internal and external links in a web page. The main idea of WSM is to discover links structure to find most commonly browsed web pages patterns (an application of WSM is to find the shortest path between two pages with high amount of visits) [11, 72]. Finally, web usage mining is the study of visitors browsing behavior when searching information in a web site. WUM is probably the most famous way of web mining, since, results allow an organization or enterprise to better understand its visitors or customers [12, 50, 58, 63, 64, 66, 71]. Other commonly used techniques are those used for user profiling (WUP). The techniques for WUP, make use of information from the visitors like text browsed, items bought, registration information, etc. this way, it is possible to create customized profiles for every visitor in a web site [14, 18, 41, 40, 42, 43, 17, 15, 51].

Standard WUM processes leave aside semantic information from web documents that may lead to poor, erroneous, hard to interpret and far from real visitors' preferences results. Some systems [17, 36] have been developed to provide good results in

---

[1] We use the term web site as the most generic way of web system. i.e. personal web site, news paper site, e-comerce sites, forums, wiki, portals, etc.

the study of visitors' browsing behavior. However, still these relay just on text similarities to understand visitors' preferences. This is why a very new class of mining techniques, which include the semantic relations from web contents have being arising (we think a new name to these techniques should be *semantic web mining*) [58]. Tools that will take advantage of all semantic information in terms of ontology or as other relations (like in social networks) which will provide web sites' analyst with high quality and valuable information to enhance a site.

In the following we are going to show several basic applications of web mining techniques to study social web sites, this is, the use of data generated on social web sites to perform data mining, without considering social aspects, i.e. mining a social web site like if it were a (static or plain) web site. Then, we are going to explain advanced techniques in web mining to study social web sites, this aims to show innovative approaches to study the social aspects in social web sites.

9.4.1 Basic Web Mining Applications to Study Social Web Sites

Even if we are processing a social web site, typical web mining questions are still valid, for example, which are the main key words in the social web site? is a WTM question; which topics are more important to the users? is a tipical WUM question; which is the better organization of hyperlinks structure in the social web site? this is a question from WSM. Also, all results of web mining process is usefull to perform

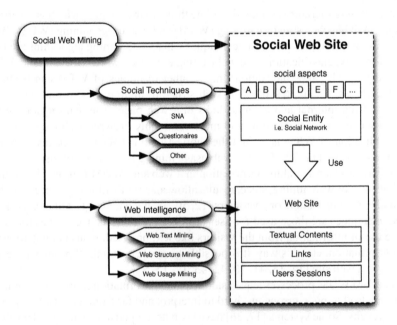

Fig. 9.1 Social Web Mining Concept

community improvement decision easier. Even if you do not know what to do, the classification would help to find the improvements that community needs.

An interesting work of WUM applied to social networks is presented in [4]. This work focuses on understanding how users behave when they connect to social networking sites. This way, authors are able to creates opportunities for better interface design, richer studies of social interactions, and improved design of content distribution systems. All this enhancements possibilities are traditionally obtained after the application of WUM into a web site. However, authors are not interested in understanding or measuring social aspects of social networks (on Table 9.1) that is why we placed this work on this section.

9.4.1.1   Automatic Classification of Posts in Social Web Sites

Sometimes, classification is used as an exploratory data tool to understand the community. However, information obtained by this tools, provide greater benefits to the community. Hong et al. [26] uses Q-A discussion boards to classify posts in question-posts or reply-posts and then find the proper reply to a question. This classification would improve the community in three ways: enhance search quality, bringing suggestions when users ask a similar question and finding experts according to the replies provided. Recognizing experts would help to solve questions faster and better in the community.

In both cases we are concern about who the experts are and how they are obtained. For that reason, this work focuses on finding a new approach to discover key members (experts or core members). The better you identifying them the more relevant will be results obtained by other tecniques. Thus, it will be possible to obtain better enhancements for the whole community.

9.4.2   Advanced Web Mining Applications to Study Social Web Sites

9.4.2.1   Finding Prestige Members with SNA and Concept-Based Text Mining

Ríos et al. [1] define the VCoP goals of a website in collaboration with community members. When goals are well established and defined, it is possible the application of concept-based text mining to evaluate the accomplishment of communities' goals. The concept-based text mining uses fuzzy logic theory to assign a "goal score" to every forum in the community. This "goal score" show how aligned to this specific goal is the text inside a community forum.

VCoP talks about specific main topics, one of his worries is that community do not deviate of them. Also, VCoPs' have purposes (or goals) that administrators want to accomplish. Therefore, it is important for VCoP to supervise the goals through time and analyse its evolution. Then, enhancements will surfaced based on this temporal purpose evolution analysis [59].

Having these scores, VCoPs' administrators evaluate the goals' accomplishment to allow administrators to make the proper enhancements. For example, if two forums have similar score of a specific goal, an enhancement could be merge both. On the other hand, if a category has very high scores in two specific goals, it is possible to split the forum in two independent forums closer to each goal.

In this work, the objective was to evaluate the goal accomplishment of VCoPs' forum, but this does not help to evaluate how users contribute to their purpose accomplishment. Thus, making difficult finding the VCoPs' key members. However, the use of concept-based mining will improve the search accuracy.

Main question of present work is how to enhance key members discovery. This question has no simple answer, the first step is to obtain a graphic representation of the inner social community. The second step is to apply an core members' algorithm (like HITS) to this representation. As a result of the algorithms' application, we will obtain a rank of all community members leaving in the top of the rank the experts (core or key) members.

We distinguish key member from expert member. Since a key member has several characteristics that define him/her. Firstly, a key member may be expert in a field or not. Secondly, he/she may increase the interaction in the community because he ask interesting questions, which produce answers from the experts on the field. This means that questions are very specific in a field, therefore, only experts are able to answer them. In other words, a key member is a person totally aligned with the VCoPs goals and topics. Thus, producing contents which are very relevant to satisfy other members interests. The only way to measure a key member as we define him is using an hybrid approach of SNA combined with semantic-based text mining. Likewise, the mining process must include always the different purposes of the community. This is why we chose the concept-based text mining approach [59].

Previous work [59] brings a method to evaluate community goals accomplishment, now we will use this approach to classify the members' posts according VCoPs' goals. These goals are defined as a set of terms, which are composed by a set of keywords or statements in natural language. To obtain a goal accomplishment score, we use fuzzy logic to evaluate how much a goal is contained in a singular post. Then, we will have a post vector in which the components will be the goals accomplishment scores of the post.

The idea is to compare with euclidean distance two members' posts and if the distance it is over a certain threshold, there will be interaction between them. We support the idea that this will help us to avoid, filter or erase irrelevant interactions. For example, in a VCoP with k goals, let $p_j^{u'}$ the post j of user u' that it is a reply to post i of user u (p_i^u). The distance between them will be calculated with Eq. (1).

$$d(p_i^u, p_j^{u'}) = \frac{\sum_k g_{ik} g_{jk}}{\sqrt{\sum_k g_{ik}^2 \sum_k g_{jk}^2}} \tag{9.1}$$

Where g_{ik} is the score of goal k in post i. It is clear that the distance exists only if $p_j^{u'}$ is a reply to p_i^u. After that, we calculate the weight of arc uu' ($w_{uu'}$) with Eq. (2).

$$w_{uu'} = \sum_{\substack{i,j \\ d(p_i^u, p_j^{u'}) \geq \theta}} d(p_i^u, p_j^{u'}) \tag{9.2}$$

We used this weight in both configurations previously described (Creator-oriented and Last Reply-oriented). Afterwards, we applied HITS to find the key members on the different networks configurations. These key members were validated with the VCoPs' administrators showing very promising results.

9.4.2.2   Measuring Social Purpose Using Concept-Based Text Mining

Although, studies of social aspects in a virtual community are crucial, the evolution analysis of such aspects is more important yet, especially in social structures that allow to its members define themselves different levels of participation among time, like Virtual Communities of Practice (VCoP) [86], provoking that community change according to member's participation.

In [59], it is proposed a methodology based in the VCoP's purpose study. The purpose refers to a community's shared focus on an information need, interest, service. Every user has his own motivations (or interests) to use a specific VCoP and of course every community of practice can satisfy specific users' needs.

Defining the community's purpose is of major importance, since potential participants can immediately find out about the communities goals [54]. However, VCoP's purpose is not always clear, even worst, it is not clear if all community members are aligned to the same common purpose.

But also, we must consider the evolutionary nature of purpose. Since, people is changing every day their motivations, interests, taste, etc. Certainly, members of a VCoP change their purpose when using the VCoP. As example, the purpose of a newcomer are to learn the basics of a theme using the VCoP. When this newcomer becomes an expert his purpose might be to research specific and complex aspects of such theme; or even more, his purpose might be to answer questions of new comers.

Therefore, our hypothesis is that study of evolutionary nature of purpose of VCoP's members is a vital way to understand a community and to enhance it. Until today there aren't significant studies to evaluate this evolution (as we have shown in section 2), and fulfillment of purpose through time. Also, we strongly support the idea that the study of social networks members' relations alone (SNA) it is not sufficient to obtain a good understanding of a VCoP. It is needed to analyze other social aspect, such as purpose, to fully understand and perform the necessary enhancements to allow VCoP exists through time.

Of course, the key aspect of this work is to consider the purpose evolution analysis as an important tool; but, how can we measure purpose? Since, purpose is something close to the ideas, or underlying motivations of every member, it is not simple to answer this question.

Since, purpose is, from dictionary, "what something is used for". We propose to use goals as a measure of purpose accomplishment. Using this idea, we can measure if a VCoP fulfills a purpose, measuring how well members' contributions accomplish a set of goals previously defined by the owners, managers or experts of the community.

Goals definition must be performed based on interviews or surveys to community experts or administrators [37, 60, 65]. Definition of goals consist of a series of phrases. These phrases respond to the question "what the community is for?".

This paper only evaluates the goals from community administrators' view point. Although, same process apply for community members' goals analysis.

Afterwards, we need to select a classification or clustering algorithm in order to perform a text mining algorithm to find interesting patterns. It is expected patterns found provide useful information for administrators and experts in order to decide how to enhance the community (add new forums, erase forums, find trends, etc) based on goals fulfillment.

We selected a concept-based text mining since the adaption from concepts to goals is straightforward. This approach will be explained in next section.

As a last consideration, this technique allows us to study the goals' fulfillment through time. Therefore, we can show interesting information of how the VCoP's purpose evolves. Thus, providing useful and objective information to community experts. Without this tool, they only have an intuition on how community has evolved and if the information contained in the community forums is truly accomplishing the purpose of the community.

9.4.2.3 Discovering and Evolution of Subgroups Using Data Mining

Communities – in the real world – are more than just the sum of participating members; we know that communities are highly dynamic social networks whose structure changes over time. In [20, 19], Falkowski et al. argue that a community not necessarily consist of the same prticipants interacting in the same way through time. Rather, members' interactions are fluctuating. Falkowski et al. propose to define a community as a persistent structure in a graph of interactions among fluctuating members.

Falkowski et al. used two approaches to study the evolution of two different types of communities. Firstly, they partition the time axis and build a graph of interactions. On this graph they apply a hierarchical divisive edge betweenness clustering algorithm to fing subgroups of densely connected individuals. Secondly, they tackle the problem of analyzing the evolution of communities in environments with a high membership fluctuation. Authors use the same clustering as in the first approach to detect subgroups in graphs (also called community instances).

To analyze subgroups Dynamics, the development of one subgroup can be described and assessed by measuring and interpreting different measures which are: stability, density and cohesion, Euclidean distance, correlation coefficient and group activity. Then, they need to develop a similarity measure to be able to compare two different community instances. This similarity uses the overlap principle. Which means that two instances are similar if they overlap exceeds a given threshold. Let t_i, t_j two different time periods and G_i, G_j the corresponding graphs of interactions and let C^{G_i}, C^{G_j} be the corresponding clusterings. Besides, $x \in C^{G_i}$ and $y \in C^{G_j}$

be two community instances. Then, Falkowski et al. define the overlap between two
community instances in Eq. (9.3). Using former equation, it was possible to define
Eq. (9.4).

$$overlap(x,y) = \frac{|x \cap y|}{\min(|x|,|y|)} \tag{9.3}$$

$$sim(x,y) = \left\{ \begin{array}{l} \tau_j - \tau_i < \tau_{period} \land \\ 1 \; overlap(x,y) \geq \tau_{overlap} \\ 0 \; otherwise \end{array} \right\} \tag{9.4}$$

Finally, Falkowski et al. implemented this methods into an interactive software and
tested the approach into real online students community.

9.5   Summary

Today, virtual communities have experienced an exponential growth. Also, the use
of web mining techniques to explode data stored in these systems has become a
natural approach to obtain knowledge from them [24, 73, 28, 44, 49, 75]. However,
a virtual community is not only a group of people accessing a web site, they establish
social relationship through the use of Internet tools [83], allowing the formation of
a communal identity and a shared sense of the world [84].

We have showed that a social entity (like social networks, communities of practice, virtual communities of practice, etc.) are characterized by social aspects. Since
there exist many types of social networks and communities, we have focused our
chapter in the study of three important social structures supported by web systems,
these are: online social networks, virtual communities and virtual communities of
practice.

We have explained all techniques to study the social aspects of these three social
entities. However, our main focus was the explanation of social network analysis
(SNA), which is commonly used to study any kind of network or community, in
particular, those supported by web systems. We have explained how SNA is applied
and results possible to be obtained by SNA. We also explained that the application
of SNA alone loses many valuable information. Therefore, the analysis of social
aspects is limited more or less depending on the type of community studied.

Afterwards, we have showed what is a social web site, and how it helps to support
these three social entities. Then we explain how web mining techniques are used to
extract valuable information from these kind of sites. However, as well as in SNA
analysis' case, the problem on the application of web mining to social web sites is
that many social aspects are not meassured.

Our main hypotheses is that to provide truly valuable information to help managers or web masters it is necessary take into account the social nature of the virtual communities in web mining techniques. Therefore, the combination of both
approaches is needed in order to trully analyze the social aspects that maintain
a community healthy/alive through time. We have called these idea, social web

mining, which is the combination of social techniques with web mining techniques to process social web sites.

We have showed several ways to combine SNA with web mining in order to process social web sites that support social networks, virtual communities and virtual communities of practice. We have discovered that, although, there are hundreds of works in the application of SNA, and hundreds more in the application of web mining techniques to study social web sites; there are quite few works in these field. Therefore, we expect that in following years more researchers see the real potential in the combination of sociological techniques like SNA with Web Mining techniques to discover more meaningfull information about human relationships to better understanding of human behavior when interacting with other people.

Acknowledgements. This work was supported partially by the Fondecyt Initiation into Reseacrh Founding project, code 11090188, entitled "Semantic Web Mining Techniques to Study Enhancements of Virtual Communities".

References

1. Alvarez, H., Ríos, S., Merlo, E., Aguilera, F., Guerrero, L.: Enhancing sna with a concept-based text mining approach to discover key members on a vcop p (to appear, 2010)
2. Arenas, A., Danon, L., Díaz-Guilera, A., Gleiser, P.: Community analysis in social networks. The European Physical Journal B-Condensed Matter (2004),
   http://www.springerlink.com/index/8R9TW62UPVXMFM9A.pdf
3. Barab, S.: Designing for virtual communities in the service of learning. The Information Society (2003), http://www.informaworld.com/index/LQTX8NCUXP3BT3QF.pdf
4. Benevenuto, F., Rodrigues, T., Cha, M.: Characterizing user behavior in online social networks. In: Proceedings of the 9th ACM SIGCOMM conference on Internet measurement conference, pp. 49–62 (2009),
   http://portal.acm.org/citation.cfm?id=1644893.1644900&
   coll=portal&dl=ACM&type=series&idx=SERIES10693&part=series&
   WantType=Proceedings&title=IMC
5. Bourhis, A., Dubé, L., Jacob, R.: The success of virtual communities of practice: The leadership factor. The Electronoc Journal of Knowledge Management $\tilde{3}$(1), 23–34 (2005),
   http://citeseerx.ist.psu.edu/viewdoc/download?doi=10.1.1.93.9460&
   rep=rep1&type=pdf
6. Breiger, R.: Duality of persons and groups, the. Soc. F (1974),
   http://heinonlinebackup.com/hol-cgi-bin/get_pdf.cgi?handle=hein.
   journals/josf53&section=26
7. Breiger, R.L., Carley, K.M., Pattison, P.: On Human Factors. N.R.C.U.C.: Dynamic social network modeling and analysis: workshop summary, vol. 2002, p. 379 (2003),
   http://books.google.com/books?id=IfnYO3YeZ_0C&printsec=frontcover

8. Breslin, J., Decker, S.: The future of social networks on the internet: The need for
   semantics. IEEE Internet Computing 11(6), 86–90 (2007),
   http://www.google.com/search?client=safari&rls=en-us&
   q=The+Future+of+Social+Networks+on+the+Internet:
   +The+Need+for+Semantics&ie=UTF-8&oe=UTF-8
9. Carotenuto, L., Etienne, W., Fontaine, M., Friedman, J., Muller, M., Newberg, H.,
   Simpson, M., Slusher, J., Stevenson, K.: Communityspace: Toward flexible support for
   voluntary knowledge communities. In: Proc. of Workshop on Workspace Models for
   Collaboration (1999),
   http://domino.watson.ibm.com/cambridge/research.nsf/0/
   0e8c8166a02d5338852568f800634af1/FILE/communityspace.PDF
10. Chakrabarti, S.: Mining the web: discovering knowledge from hypertext data. Part 2,
    345 (2003),
    http://books.google.com/books?id=5Zxw1h6yc_UC&printsec=frontcover
11. Chakrabarti, S., Dom, B., Gibson, D., Kleinberg, J.: Mining the link structure of the
    world wide web. IEEE Computer (1999),
    http://www.cs.cornell.edu/home/kleinber/ieee99-web.pdf
12. Cooley, R., Mobasher, B., Srivastava.., J.: Data preparation for mining world wide web
    browsing patterns. Knowledge and Information Systems 1(1) (1999),
    http://maya.cs.depaul.edu/~mobasher/papers/webminer-kais.pdf
13. Craven, P., Wellman, B.: The network city, p. 126 (1973),
    http://books.google.com/books?id=MCSLGAAACAAJ&printsec=frontcover
14. Dai, H., Mobasher, B.: A road map to more effective web personalization: Integrating
    domain knowledge with web usage mining. In: Proceedings of the International Confer-
    ence on Internet Computing 2003 (2003),
    http://maya.cs.depaul.edu/~mobasher/papers/DM03.pdf
15. Datta, A., Dutta, K., VanderMeer, D., Ramamritham, K.: An architecture to support scal-
    able online personalization on the web. The VLDB Journal (2001),
    http://www.springerlink.com/index/8W4MV36RT5D1BB42.pdf
16. Ehrlich, K., Lin, C., Griffiths-Fisher, V.: Searching for experts in the enterprise: com-
    bining text and social network analysis. In: Proceedings of the 2007 international ACM
    conference on Supporting group work (2007),
    http://portal.acm.org/citation.cfm?id=1316642
17. Eirinaki, M., Vazirgiannis, M.: Web mining for web personalization. ACM Transactions
    on Internet Technology, TOIT (2003),
    http://portal.acm.org/citation.cfm?id=643478
18. Eirinaki, M., Vazirgiannis, M., Varlamis, I.: Sewep: using site semantics and a taxonomy
    to enhance the web personalization process. In: Proceedings of the 9th ACM SIGKDD
    International Conference on Knowledge Discovery and Data Mining (2003),
    http://portal.acm.org/citation.cfm?id=956765
19. Falkowski, T., Bartelheimer, J., Spiliopoulou, M.: Mining and visualizing the evolution
    of subgroups in social networks. In: WI 2006: Proceedings of the 2006 IEEE/WIC/ACM
    International Conference on Web Intelligence (2006),
    http://portal.acm.org/citation.cfm?id=1248823.1249048
20. Falkowski, T., Spiliopoulou, M.: Data mining for community dynamics. Künstliche
    Intelligenz (2007),
    http://www.kuenstliche-intelligenz.de/fileadmin/template/main/
    archiv/pdf/ki2007-03_page23-29_web_full.pdf

21. Fetterman, D.: Ethnography: Step by step. orton.catie.ac.cr (1998),
    `http://orton.catie.ac.cr/cgi-bin/wxis.exe/?IsisScript=SIBE01.xis&`
    `method=post&formato=2&cantidad=1&expresion=mfn=034353`
22. Fredericks, K., Durland, M.: The historical evolution and basic concepts of social net-
    work analysis. New directions for evaluation (2006),
    `http://www3.interscience.wiley.com/journal/112391275/abstract`
23. Garton, L., Haythornthwaite, C., Wellman, B.: Studying online social networks. Journal
    of Computer-Mediated Communications 3(1) (1997),
    `http://www.google.com/search?client=safari&rls=en-us&`
    `q=Studying+online+social+networks&ie=UTF-8&oe=UTF-8`
24. Géry, M., Haddad, H.: Evaluation of web usage mining approaches for user's next request
    prediction. In: Proceedings of the 5th ACM international workshop on Web information
    and data management (2003),
    `http://portal.acm.org/citation.cfm?id=956716`
25. Henri, F., Pudelko, B.: Understanding and analysing activity and learning in virtual com-
    munities. Journal of Computer Assisted Learning 19, 474–487 (2003),
    `http://hal.archives-ouvertes.fr/hal-00190267/`
26. Hong, L., Davison, B.: A classification-based approach to question answering in dis-
    cussion boards. In: Proceedings of the 32nd international ACM SIGIR conference on
    Research and development in information retrieval, pp. 171–178 (2009),
    `http://www.springerlink.com/content/l717p16267004158/`
27. Huang, W., Hong, S.H., Eades, P.: How people read sociograms: a questionnaire study.
    In: APVis 2006: Proceedings of the 2006 Asia-Pacific Symposium on Information Visu-
    alisation, vol. 60 (2006),
    `http://portal.acm.org/citation.cfm?id=1151903.1151932`
28. Jin, X., Zhou, Y., Mobasher, B.: Web usage mining based on probabilistic latent seman-
    tic analysis. In: Proceedings of the tenth ACM SIGKDD international conference on
    Knowledge discovery and data mining (2004),
    `http://portal.acm.org/citation.cfm?id=1014052.1014076`
29. Johnson, C.: A survey of current research on online communities of practice, The internet
    and higher education. Elsevier, Amsterdam (2001),
    `http://linkinghub.elsevier.com/retrieve/pii/S1096751601000471`
30. Jones, Q.: Virtual-communities, virtual settlements & cyber-archaeology: A theoretical
    outline. Journal of Computer Mediated Communication (1997),
    `http://jcmc.indiana.edu/vol3/issue3/jones.html?`
    `ref=totalcasinoguide.info`
31. Kim, A.J.: Community Building on the Web: Secret Strategies for Successful Online
    Communities (2000),
    `http://www.amazon.com/exec/obidos/redirect?tag=citeulike07-20%5C&`
    `path=ASIN/0201874849`
32. Kim, W., Jeong, O.R., Lee, S.W.: On social web sites. Information Systems 35(2),
    215–236 (2010),
    `http://www.sciencedirect.com/science/article/B6V0G-4X5YT84-1/%2/`
    `84a742613988ab4ee8b8f2ff0bd7ae54`
33. Koh, J., Kim, Y., Butler, B., Bock, G.: Encouraging participation in virtual communities.
    Communications of the ACM (2007),
    `http://portal.acm.org/citation.cfm?id=1216016.1216023`
34. Kosala, R., Blockeel, H.: Web mining research: a survey. ACM SIGKDD Explorations
    Newsletter (2000),
    `http://portal.acm.org/citation.cfm?id=360406`

35. Kozinets, R.: The field behind the screen: Using netnography for marketing research in
    online communities. Journal of Marketing Research (2002),
    http://www.atypon-link.com/AMA/doi/abs/10.1509/jmkr.39.1.61.18935
36. Liu, H.: Kešelj, V.: Combined mining of web server logs and web contents for classifying
    user navigation patterns and predicting users' future requests. In: Data & Knowledge
    Engineering (2007),
    http://portal.acm.org/citation.cfm?id=1231807
37. Loh, S., de Oliveira, J., Gameiro, M.: Knowledge discovery in texts for constructing
    decision support systems. Applied Intelligence (2003),
    http://www.springerlink.com/index/L7L5787M0J82J0WN.pdf
38. Marathe, J.: Creating community online. Durlacher Research Ltd. (1999),
    http://www.delijst.net/delijst/pdf/unpan003006.pdf
39. Marsden, P., Lin, N.: Social structure and network analysis (1982),
    http://en.scientificcommons.org/22431038
40. Mobasher, B., Cooley, R., Srivastava, J.: Automatic personalization based on web usage
    mining. Communications of the ACM (2000),
    http://portal.acm.org/citation.cfm?doid=345124.345169
41. Mobasher, B., Dai, H., Luo, T., Nakagawa, M.: Effective personalization based on as-
    sociation rule discovery from web usage data. In: Workshop On Web Information And
    Data Management (2001),
    http://portal.acm.org/citation.cfm?id=502932.502935
42. Mobasher, B., Dai, H., Luo, T., Sun, Y., Zhu, J.: Integrating web usage and content
    mining for more effective personalization. In: Bauknecht, K., Madria, S.K., Pernul, G.
    (eds.) EC-Web 2000. LNCS, vol. 1875, pp. 165–176. Springer, Heidelberg (2000),
    http://books.google.com/books?hl=en&lr=&ie=UTF-8&id=kb69hBiQMiYC&
    oi=fnd&pg=PA165&dq=fbGsqm_a8JcJ:scholar.google.com/&ots=6WCTl9IzVr&
    sig=bMwSGS73lyQlzN6cVewJp7dlNg8
43. Mulvenna, M., Anand, S., Büchner, A.: Personalization on the net using web mining:
    Introduction. Communications of the ACM (2000),
    http://portal.acm.org/citation.cfm?doid=345124.345165
44. Nasraoui, O., Soliman, M., Saka, E., Badia, A., Germain, R.: A web usage mining frame-
    work for mining evolving user profiles in dynamic web sites. IEEE Transactions on
    Knowledge and Data Engineering (2008),
    http://doi.ieeecomputersociety.org/10.1109/TKDE.2007.190667
45. Nocera, J.: Ethnography and hermeneutics in cybercultural research accessing irc virtual
    communities. Journal of Computer-Mediated Communication (2002),
    http://www.blackwell-synergy.com/doi/abs/10.1111/j.1083-6101.2002.
    tb00146.x
46. Pal, S., Talwar, V., Mitra, P.: Web mining in soft computing framework: relevance, state
    of the art and future directions. Neural Networks (2002),
    http://ieeexplore.ieee.org/xpls/abs_all.jsp?arnumber=1031947
47. Perkowitz, M.: Adaptive web sites: Cluster mining and conceptual clustering for index
    page synthesis. perkowitz.net (2001),
    http://www.perkowitz.net/research/papers/phd.ps.gz
48. Perkowitz, M., Etzioni, O.: Towards adaptive web sites: Conceptual framework and case
    study. Artificial Intelligence (2000),
    http://linkinghub.elsevier.com/retrieve/pii/S0004370299000983

49. Perugini, S., Goncalves, M., Fox, E.: Recommender systems research: A connection-centric survey (2004),
    `http://apps.isiknowledge.com/InboundService.do?Func=Frame&`
    `product=WOS&action=retrieve&SrcApp=Papers&UT=000223535400001&`
    `SID=1BbJAfa4gIa8KKFIpkl&Init=Yes&SrcAuth=mekentosj&mode=FullRecord&`
    `customersID=mekentosj&DestFail=http%253A%252F%252Faccess.`
    `isiproducts.com%252Fcustom_images%252Fwok_failed_auth.html`
50. Pierrakos, D., Paliouras, G., Papatheodorou, C.: Web usage mining as a tool for personalization: A survey. User Modeling and User-Adapted Interaction (2003),
    `http://www.springerlink.com/index/X0T6WPPW58883587.pdf`
51. Pierrakos, D., Paliouras, G., Papatheodorou, C., Spyropoulos, C.: Web usage mining as a tool for personalization: A survey. User Modeling and User-Adapted Interaction 13(4), 311–372 (2003)
52. Plaskoff, J.: Intersubjectivity and community building: Learning to learn organizationally, pp. 161–184 (2003)
53. Preece, J.: Sociability and usability in online communities: determining and measuring success. Behaviour & Information Technology (2001),
    `http://www.informaworld.com/index/M9EMFTN4DGR0DAPA.pdf`
54. Preece, J.: Etiquette, empathy and trust in communities of practice: Stepping-stones to social capital. Journal of Universal Computer Science (2004),
    `http://www.jucs.org/jucs_10_3/etiquette_empathy_and_trust/Preece_J.`
    `html`
55. Preece, J., Maloney-Krichmar, D.: Online communities: Focusing on sociability and usability. In: Handbook of Human-Computer Interaction (2003),
    `http://isis.ku.dk/kurser/blob.aspx?feltid=102191`
56. Probst, G., Borzillo, S.: Why communities of practice succeed and why they fail. European Management Journal 26(5), 335–347 (2008),
    `http://www.sciencedirect.com/science/article/B6V9T-4SWP24X-1/2/`
    `dbb451298682776766f494b7e25154e6`
57. Rheingold, H.: A slice of my life in my virtual community. Global networks: Computers and international communication, 57–80 (1993),
    `http://books.google.com/books?hl=en&lr=&id=xI_Um3dTTeYC&oi=fnd&`
    `pg=PA413&dq=A+Slice+of+Life+in+My+Virtual+Community&ots=iUN7YtlQ8n&`
    `sig=peOI5xd3N6hGvmFM-h2obW9jf6A`
58. Ríos, S.A.: A study on web mining techniques for off-line enhancements of web sites. Ph.D Thesis p. 231 (2007)
59. Ríos, S.A., Aguilera, F., Guerrero, L.: Virtual communities of practice's purpose evolution analysis using a concept-based mining approach. Knowledge-Based and Intelligent Information and Engineering Systems 2, 480–489 (2009)
60. Ríos, S.A., Velásquez, J.D.: Semantic web usage mining by a concept-based approach for off-line web site enhancements. In: 2008 IEEE/WIC/ACM International Conference on Web Intelligence and Intelligent Agent Technology, vol. 1, pp. 234–241 (2008), doi:10.1109/WIIAT.2008.406,
    `http://ieeexplore.ieee.org/search/srchabstract.jsp?`
    `arnumber=4740455&isnumber=4740405&punumber=4740404&`
    `k2dockey=4740455ieeecnfs`
61. Ríos, S.A., Velásquez, J.D., Vera, E.S., Yasuda, H., Aoki, T.: Using sofm to improve web site text content. In: Wang, L., Chen, K., S. Ong, Y. (eds.) ICNC 2005. LNCS, vol. 3611, pp. 622–626. Springer, Heidelberg (2005),
    `http://www.springerlink.com/index/dmt1u7rld84cv2mr.pdf`

62. Ríos, S.A., Velásquez, J.D., Yasuda, H., Aoki, T.: Web site improvements based on representative pages identification. In: Zhang, S., Jarvis, R.A. (eds.) AI 2005. LNCS (LNAI), vol. 3809, pp. 1162–1166. Springer, Heidelberg (2005),
http://www.springerlink.com/index/pp1125r774w3358m.pdf

63. Ríos, S.A., Velásquez, J.D., Yasuda, H., Aoki, T.: Conceptual classification to improve a web site content. In: Corchado, E., Yin, H., Botti, V., Fyfe, C. (eds.) IDEAL 2006. LNCS, vol. 4224, pp. 869–877. Springer, Heidelberg (2006)

64. Ríos, S.A., Velásquez, J.D., Yasuda, H., Aoki, T.: A hybrid system for concept-based web usage mining. International Journal of Hybrid Intelligent Systems (IJHIS) 3(4), 219–235 (2006),
http://iospress.metapress.com/index/6JB1PJ9VVF5F0TW0.pdf

65. Ríos, S.A., Velásquez, J.D., Yasuda, H., Aoki, T.: Using a self organizing feature map for extracting representative web pages from a web site. International Journal of Computational Intelligence Research (IJCIR) 2, 159–167 (2006),
http://www.softcomputing.net/ijcir/1003a.pdf

66. Ríos, S.A., Velásquez, J.D., Yasuda, H., Aoki, T.: Web site off-line structure reconfiguration: A web user browsing analysis. In: Gabrys, B., Howlett, R.J., Jain, L.C. (eds.) KES 2006. LNCS (LNAI), vol. 4252, pp. 371–378. Springer, Heidelberg (2006),
http://www.springerlink.com/content/k1147742h457/

67. Rogers, E.M., Kincaid, D.L.: Communication networks: Toward a new paradigm for research, p. 386,
http://www.amazon.com/
Communication-Networks-Toward-Paradigm-Research/dp/0029267404

68. Scime, A.: Web mining: applications and techniques? p. 427 (2005),
http://books.google.com/books?id=TDhPMs3adw0C&printsec=frontcover

69. Scott, J.: Social network analysis: a handbook, p. 208 (2000),
http://books.google.com/books?id=Ww3_bKcz6kgC&printsec=frontcover

70. Shummer, T.: Patterns for building communities in collaborative systems. In: Proceedings of the 9th European Conference on Pattern Languages and Programs (2004),
http://hillside.net/europlop/europlop2004/Papers/wwc/C5.pdf

71. Spiliopoulou, M.: Web usage mining for web site evaluation. Communications of the ACM (2000),
http://portal.acm.org/citation.cfm?doid=345124.345167

72. Spiliopoulou, M., Mobasher, B., Berendt, B., Nakagawa, M.: A framework for the evaluation of session reconstruction heuristics in web-usage analysis. INFORMS Journal on Computing (2003),
http://warhol.wiwi.hu-berlin.de/~berendt/Papers/spiliopoulou_etal_2003.pdf

73. Srivastava, J., Cooley, R., Deshpande, M., Tan, P.: Web usage mining: discovery and applications of usage patterns from web data. In: ACM SIGKDD Explorations Newsletter (2000),
http://portal.acm.org/citation.cfm?id=846188&dl=GUIDE

74. Sudweeks, F., Simoff, S.J.: Complementary explorative data analysis: the reconciliation of quantitative and qualitative principles. Doing Internet Research: Critical Issues and Methods for Examining the Net, 29–55 (1999)

75. Tao, Y., Hong, T., Su, Y.: Web usage mining with intentional browsing data. Expert Systems With Applications (2008),
http://linkinghub.elsevier.com/retrieve/pii/S0957417407000668

76. Turney, P.: Mining the web for lexical knowledge to improve keyphrase extraction: Learning from labeled and unlabeled data. Arxiv preprint cs.LG (2002),
    https://iit-iti.nrc-cnrc.gc.ca/iit-publications-iti/docs/
    NRC-44947.pdf
77. Wasserman, S., Faust, K.: Social network analysis: Methods and applications. books.google.com (1994),
    http://books.google.com/books?hl=en&lr=&id=CAm2DpIqRUIC&oi=fnd&
    pg=PR21&dq=clustering+social+networks&ots=HtMntfXEPe&sig=VaA_
    -1yEkg_euBpXsJOCcnCclRk
78. Wellman, B.: An electronic group is virtually a social network. Culture of the Internet (1997),
    http://books.google.com/books?hl=en&lr=&id=5uarXm1CkccC&oi=fnd&
    pg=PA179&dq=%2522An+Electronic+Group+is+Virtually+a+Social+
    Network%2522&ots=KS6xYR3g4S&sig=kl-Gh63xd_Qsg_MrUOGMYez4pDg
79. Wellman, B.: Changing connectivity: A future history of y2. 03k. socresonline.org.uk (2000),
    http://www.socresonline.org.uk/cgi-bin/perlfect/search/search.pl?
    q=chris&showurl=%252F4%252F4%252Fwellman.html
80. Wellman, B.: Computer networks as social networks. Science 293, 2031–2035 (2001),
    http://adsabs.harvard.edu/abs/2001Sci...293.2031W
81. Wellman, B., Berkowitz, S.D.: Social structures: A network approach, vol. 15, p. 528. Emerald Group Publishing Limited (1998)
82. Wellman, B., Gulia, M.: Virtual communities as communities. Communities in cyberspace (1999),
    http://books.google.com/books?hl=en&lr=&id=harO_jeoyUwC&oi=fn%d&
    pg=PA167&dq=%2522Virtual+communities+as+communities%2522&
    ots=JWUI8GaxsS&sig=FIcftIzZwMS1GfCHTxokSxFdjBg
83. Wellman, B., Salaff, J., Dimitrova, D., Garton, L.: Computer networks as social networks: Collaborative work, telework, and virtual community. Annual Reviews in Sociology (1996),
    http://arjournals.annualreviews.org/doi/abs/10.1146%252Fannurev.
    soc.22.1.213
84. Wenger, E.: Communities of practice: Learning, meaning, and identity. books.google.com (1999),
    http://books.google.com/books?hl=en&lr=&ie=UTF-8&id=heBZpgYUK%dAC&
    oi=fnd&pg=PR11&dq=%2522Wenger%2522&ots=kcmh-sbxZk&sig=8F_
    tXXNLuHe4r5FEE42t%cVcmoxU
85. Wenger, E.: Communities of practice: Learning, meaning, and identity (1999)
86. Wenger, E., McDermott, R., Snyder, W.: Cultivating communities of practice: A guide to managing knowledge. Harvard Business School Press, Boston (2002),
    http://emergence.org/ECO_site/web-content/V8_Books/details/1095.
    html
87. Wenger, E., McDermott, R.A., Snyder, W.: Cultivating communities of practice (2002)
88. Wenger, E., White, N., Smith, J.: Digital habitats; stewarding technology for communities. books.google.com (2010),
    http://books.google.com/books?hl=en&lr=&id=E7GPhmV4-KkC&oi=fnd&
    pg=PR11&dq=%2522Technology+for+communities%2522&ots=2BXXiV7Aux&
    sig=zH9dDtb-In4RU24FU9hYOt3AsfM

89. Wenger, E., White, N., Smith, J., Rowe, K.: Technology for communities. CEFRIO
    Book Chapter (January 2005),
    http://waterwiki.net/images/9/97/Technology_for_communities_-_
    book_chapter.pdf
90. Xu, J., Chen, H.: Crimenet explorer: a framework for criminal network knowledge dis-
    covery. In: ACM Transactions on Information Systems, TOIS (2005),
    http://portal.acm.org/citation.cfm?id=1059984 (Aqui sale bien el block-
    modeling)

Chapter 10
Intelligent Ubiquitous Services Based on Social Networks

Jason J. Jung

Abstract. A number of studies have been conducted on discovering useful information from social networks among people. Particularly, on ubiquitous environment, the social network between people, regarded as the channel for exchanging and propagating their contexts, plays a crucial role on being aware of the user contexts. To efficiently discover the contexts of a certain users, the contexts of his neighbors on the social network can be fused to provide mobile recommendation services to mobile subscribers. In this chapter, we focus on a social network-based service provision on the ubiquitous environment, and introduce two case studies. Firstly, mobile services have been provided by discovering social networks. Telecommunication data collected from mobile subscribers has been statistically analyzed. Secondly, all possible on- and off-line social networks were mobilized to build an ego-centric social network, since the social network of a user has been fragmented into more than one. We have collected the social network dataset from online (e.g., Facebook, Twitter, CyWorld, and co-authoring patterns in major Korean journals) and offline (e.g., co-participation patterns in a number of Korean domestic conferences). Once the ego-centric social networks have been configured, mobile services were provided to the conference participants by sending text messages about time schedule of relevant presentations. In conclusion, the social network among mobile users can play an important role in determining whether the ubiquitous services is contextually relevant or not.

10.1 Introduction

In order to efficiently provide relevant services to mobile users, it is important for service providers to gather as much useful information related to user contexts as possible (e.g., where the users are located), and to find out their "contexts"

Jason J. Jung
Knowledge Engineering Laboratory, Yeungnam University, Gyeongsan, Republic of Korea
e-mail: j2jung@(intelligent.pe.kr,ynu.ac.kr,gmail.com)

J.D. Velásquez and L.C. Jain (Eds.): Advanced Techniques in Web Intelligence – 1, SCI 311, pp. 251–264.
springerlink.com
© Springer-Verlag Berlin Heidelberg 2010

(e.g., what they are looking for). However, service providers still have some difficulties on being aware of *personal context* at a certain moment and place, because
of a large number of uncertainties. The uncertain and unpredictable factors can be
dealt with by several *contextual fusion* approaches [1]. This approach aims to the integration of as many contexts as possible. And has been applied in various domains
such as location-based systems [2, 3] as well as multiple expert systems [4], image
processing by multitemporal contextual information [5] and information retrieval
systems [6, 7].

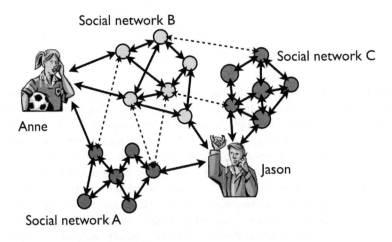

Fig. 10.1 Mobile services based on social networks

One of the latest studies for such a context-awareness process is to take into
account contextual dependency in a social network [8].

Definition 10.1 (Social network). A social network *SN* is given by

$$SN = \langle U, N, C_N \rangle \qquad (10.1)$$

where U is a set of users in the social network, and $N \subseteq U \times U$ indicating social links
between users. Also C_N means a social context of the corresponding link between
two users.

Particularly, in this chapter, *social* (or social affinity) context C_N (e.g., where users
are socially related with each other) extracted from a social network among the
mobile users is the main target out of many types of contexts. The personal context
of an user is significantly influenced by socially-related people, and vice versa [9,
10]. Hence, it is referred to as a social context in this chapter.

For example, as shown in Fig. 10.1, if Jason and Anne have a certain social
relationship (e.g., family, friend, teammate, and colleague), then their contexts are
mutually interdependent with each other. In Table 10.1, when Jason is in a restaurant

(i.e., Case$_1$ and Case$_2$), his personal contexts in two cases (i.e., CTX$_1$ and CTX$_2$) might differ from each other, depending on which guests he is with (e.g., his father or his friend).

Table 10.1 An example of context dependencies of Jason

Contexts	Spatial context	Social affinity context	Integrated context
Case$_1$	Restaurant	Friend	CTX$_1$
Case$_2$	Restaurant	Father	CTX$_2$
Case$_3$	Hospital	Father	CTX$_3$

Of course, similar to the previous context, this social context is also able to have some dependencies on other contexts. For example, in two cases Case$_2$ and Case$_3$, his personal contexts CTX$_2$ and CTX$_3$ are not identical, because the social context might be depended on the spatial contexts (i.e., in "Restaurant" or "Hospital").

In order to formulate the social contexts and contextual dependencies between the social contexts, we would live to demonstrate two possible approaches by introducing two case studies.

- Recommendation by building social networks
- Recommendation by using external social networks

Regarding the first approach, several studies have been carried out to elicit the social networks from a large number of dataset from telecommunication companies. In particular, the so-called "reality mining" [11] has introduced some possible scenarios by using data mining and visualization technologies. A number of types of usage patterns collected from mobile environments have been analyzed to figure out the social relationships between two arbitrary users. Particularly, in our previous work [8], an interactive approach has been proposed to discover the social networks between the users.

The motivation of the second case is to solve two main drawbacks on exploiting data mining tools to discover the social networks and to provide mobile users with only restricted services.

Privacy issues.   As the first problem, difficulties were encountered on protecting the privacy of the users. Even if they have allowed the company to use their data, we were not able to justify whether their social networks discovered by the data mining tools are correct or not without real answers from the users. It is impossible for the users to confirm the discovered social networks.

Isolation of social networks.   The second problem which is even more serious is that the discovered social network is isolated (or fragmented). In practice, telecommunication companies have never tried to share their subscriber data with other companies. Thus, it mean the real dataset obtained from one of the companies can reflex only a subset of the social network that one wishes to find.

As shown in Fig. 10.1, even though Jason is Anne's friend, the social context between them can be propagated.

More importantly, a context fusion process which is to integrate the collected context information should be effectively conducted for making better decisions.

In order to solve these problems, we focused on importing the social networks which already exist in other systems, instead of discovering social networks by the data mining tools. Thus, one can expect that the chance of discovering the social network between people who are using different telecommunication companies will be increased. Again, as previously mentioned all possible contexts should be fused, as shown in Fig. 10.1, the social networks imported from other systems should be efficiently integrated with each other.One refers this process as the *mobilization* of social networks . Therefore, a social network of each user, called an *ego-centric* social network , has been constructed for the purpose of efficient service provision.

Definition 10.2 (An ego-centric social network). An ego-centric social network of user u can be simply given by

$$SN_u^{\tilde{\varepsilon}} = \{SN_i|u \in SN_i\}. \tag{10.2}$$

It indicates a power set of social networks the user participates in. There should be a user identification to justify if the neighbors (i.e., u' and u'') in different social networks are identical (i.e., $u,u' \in SN_i$, $u,u'' \in SN_j$). In this work, we have done it manually.

Now, there can be a number of possibilities which provide mobile services, depending on the strategies and marketing policies of businesses (e.g., advertisement). In this chapter, some personalized mobile services are shown based on context matching between users.

The outline of this chapter is as follows. In Sect. 10.2, the first case is explained for discovering an interactive social network from telecommunication usage patterns. Sect. 10.3 addresses an ontology-based context fusion method for heterogeneous contexts extracted from multiple social networks. Sect. 10.4 shows how to generate recommendation for each user as the mobile service. Also, there are some possible scenarios for the recommendation services. In Sect. 10.5 and Sect. 10.6, a case study has been conducted in local conferences and several implications have been found out through the case study presented, respectively. Finally, a conclusion is drawn of this work in Sect. 10.7.

10.2 Interactive Discovery of Social Networks

We investigated the interactive discovery process for analyzing a large amount of datasets including usage patterns collected from mobile users, as shown in Fig. 10.2. The main goal of this discovery process is to justify whether there is a certain social relationship between two arbitrary users, and to enrich social propositions to a social network by referring to the social network ontology (SNO) . The SNO designed by human experts can include the background knowledge about the social relationships. Sometimes, it is necessary to merge several ontologies into a consensual ontology for removing redundancy and inconsistency problems. (In next section, the ontologies is explained in more detail.)

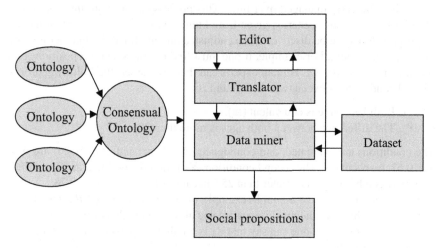

Fig. 10.2 Interactive discovery process for social propositions

The interactive discovery process is simply composed of the following four steps;

- Step 1: A human expert can input a propositional sentence based on his experiences without any quantitative modifier to the system. He can use a GUI tool (i.e., the editor module in Fig. 10.2).
- Step 2: The proposition can be easily translated into a mathematical algebra by the traslator module in Fig. 10.2. In fact, they are just symbols to be compared.
- Step 3: A data miner module can scan the collected dataset to measure the confidence level of the proposition given by the expert.
- Step 4: Through the GUI tool, the confidence level of each proposition can be shown to the human expert.

These process should be repeated until the best combination of conditions is found. The mobile usage pattern datasets have been sampled from KTF[1] legacy databases where raw records are stored. Mainly, it consists of three parts: (1) registration profiles, (2) device (and service) specifications, and (3) the calling patterns of over 60,000 mobile users. These datasets are applied to predict social contexts between mobile users by the interactive discovery process. We attempted to formalize the scenarios which are easily understandable. Thus, each scenario can be matched to a set of social contexts. Given a certain social context, we have investigated as many cases (i.e., social propositions) as possible. At the initial stage, we have

---

[1] KTF is the second largest telecommunication company in Korea.

interviewed with domain experts in KTF. For data miner module, in our case, we have exploited machine learning software packages (e.g., SPSS Clementine [http://www.spss.com/clementine/] and Weka [http://www.cs.waikato.ac.nz/ml/weak/]).

When repeating the discovery process, the number of social propositions can increase. Eventually, a number of the social propositions are represented as a decision tree to be efficiently managed over time. This process can find out the best orders of fields to verify any scenarios given from human experts. We believe that the social propositions can be discovered (i.e., adjusted and modified) by interacting with the proposed system. For example, if one must find out the social propositions of a social context *isFatherOf*. Human experts can therefore assert the following propositional conditions between two users A and B;

- $\mathscr{P}_1$: Both last name is equivalent ($\equiv$).
- $\mathscr{P}_2$: The difference between both ages is more than 20.

The conditions are evaluated, and confidence $\mathscr{L}(\mathscr{P}_1, \equiv)$ and $\mathscr{L}(\mathscr{P}_2, 20)$ are 0.99 and 0.67, respectively. The second condition $\mathscr{P}_2$ might be modified as "The difference between both ages is greater than 25," because of $\mathscr{L}(\mathscr{P}_2, 25) = 0.78$.

Here, one aims to give a simple example of social relation *isFatherOf* and *isFamilyWith*, which are the most important social relation in this project. At the first step, by common sense, one can say that n_A is a father of n_B when either P_1 or P_2 in Table 10.2 is satisfied.

Table 10.2 Social propositions by common sense about *isFatherOf*

$$
\begin{array}{l}
P_1:\ (Payment(n_B) = n_A) \\
\quad \wedge(Age(n_A) - Age(n_B) \in [30, 50]) \\
\quad \wedge(Lastname(n_B) = Lastname(n_A)) \\
P_2:\ (Location(n_A, AtNight) = Location(n_B, AtNight)) \\
\quad \wedge(Age(n_A) - Age(n_B) \in [30, 50]) \\
\quad \wedge(Lastname(n_B) = Lastname(n_A))
\end{array}
$$

When interacting with the proposed data mining module, the real dataset can be analyzed to annotate each proposition with statistical confidence. Thus, the proposition from the first step can be branched to more specific propositions, as shown in Table 10.3.

In addition, this logical expressions can be dynamically updated over time. More importantly, the social relations are semantically represented as concept hierarchy. For example, *isFatherOf* and *isBrotherOf* are subclasses of *isFamilyWith*. Thus, when the given information is not clear or limited, one can replace it to one of its superclass relations.

Table 10.3 Adjusted Social propositions by interactions about *isFatherOf*

P_{1-1}: $(Payment(n_B) = n_A, 60\%)$
$\quad \wedge (Age(n_A) - Age(n_B) \in [30,40], 85\%)$
$\quad \wedge (Lastname(n_B) = Lastname(n_A))$
P_{1-2}: $(Payment(n_B) = n_A, 60\%)$
$\quad \wedge (Age(n_A) - Age(n_B) \in [40,50], 60\%)$
$\quad \wedge (Lastname(n_B) = Lastname(n_A))$
P_2: $(Location(n_A, AtNight) = Location(n_B, AtNight))$
$\quad \wedge (Age(n_A) - Age(n_B) \in [30,50])$
$\quad \wedge (Lastname(n_B) = Lastname(n_A))$

10.3 Ontology-Based Context Fusion

Not only discovering a social network is an important issue but also reusing and integrating the existing social networks. In this section, we aim to address a context fusion method to integrate contextual evidences that have been collected from the social networks. However, the semantics in a social network is different from others by nature. Such heterogeneity of semantics not only includes the linguistic differences (e.g., between 'reference' and 'bibliography') but also mismatching between conceptual structures. In order to deal with the problems, an ontology is exploited from each social network, with an ontology matching method.

Definition 10.3 (Ontology). An ontology of a social network *SN* can be simplified by

$$\mathscr{O}_{SN} = \langle \mathscr{C}, \mathscr{P}, \mathscr{I}, \mathscr{R}_{\mathscr{C}}, \mathscr{R}_{\mathscr{P}} \rangle \tag{10.3}$$

where $\mathscr{C}$, $\mathscr{P}$, and $\mathscr{I}$ are the sets of concepts, properties, and instances contained in *SN*, respectively. Also, $\mathscr{R}_{\mathscr{C}} \subseteq \mathscr{C} \times \mathscr{C}$, and $\mathscr{R}_{\mathscr{P}} \subseteq \mathscr{P} \times \mathscr{P}$.

For example, in our previous work [8], a social network ontology (SNO) has been constructed for representing family relationships (e.g., 'isFatherOf', 'isMotherOf', and so on). Recently, various studies about SNO have been investigated (e.g., [12, 13]). More importantly, one realizes that the semantics of social networks have been already utilized from other domains. For example, DBLP[2] ontology[3] is an ontology for online bibliographic database system. From here, we can extract various social relationships (e.g., 'coauthor'), which is not interoperable with a SNO from LinkedIn , as shown in Fig. 10.3 .

One is able to apply OWL ontology matching facilities to solve this problem. Practically, such ontologies are already written in OWL[4] . Once ontology matching process has been carried out by using software tools [14], one can obtain a set of correspondences between ontology entities (denoted as *CORR*). For example,

---

[2] http://dblp.uni-trier.de/

[3] http://swat.cse.lehigh.edu/resources/onto/dblp.owl

[4] Web Ontology Language (OWL). http://www.w3.org/TR/owl-features/

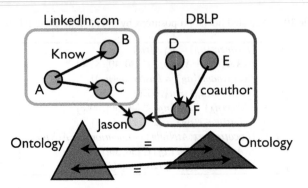

Fig. 10.3 Heterogeneous social network ontologies; $SN_{\mathrm{DBLP}} = \{D, E, F, Jason\}$ and $SN_{\mathrm{LinkedIn}} = \{A, B, C, Jason\}$

DBLP:coauthor $\sqsubseteq$ LinkedIn:Know in Fig. 10.3. (We do not aim to explain more descriptions about ontology matching in details.)

One has to take into account how to represent the user contexts, and how to propagate the contexts along the social networks. So far, there have many kinds of definitions for these contexts [15]. The contexts of a user are considered as both personal (e.g., user preferences and interests) and environmental (e.g., temperature and so on). In this study, we focus on the following two contexts of a certain user, i.e., research topics ($CTX^{\mathscr{T}}$) and current location ($CTX^{\mathscr{L}}$).

Additionally, we claim that the contexts can be propagated to other users on the social networks. It means social networks among users are playing the role of a channel where the context information can be transferred.

Definition 10.4 (Social context). Given a user u, his social context can be represented as a set of properties and their scores

$$CTX^{\mathscr{L}}_{u \to u'} = \{ \langle p_i, scr_i \rangle | p_i \in \mathscr{O}_{SN_i}, SN_i \in SN^{\tilde{\mathscr{E}}}_u \} \tag{10.4}$$

where $\mathscr{O}_{SN_i}$ is the SNO in the corresponding user's ego-centric network $SN^{\tilde{\mathscr{E}}}_u$. Also, one can simplify the score of each property to $scr = 1$, if there is a social relationship between u and u'. Even though there is no direct link between two users, the social context can be efficiently aggregated to the shortest path between both users by ontological reasoning [16].

More importantly, depending on the social context of the link (or path) between two users, neighbors' contexts can be transformed (or triggered). Practically, the score is set to 1 if the social context is positive. At the moment, only positive social contexts are considered. For example, if the relationship between two users Jason and Anne is positive (e.g., friends and colleague) in Fig. 10.1, then the context of Jason should be *influenced* to that of Anne (and vice versa).

Now, we aim to present two main heuristics to determine whether (and how much) a social context is positive or doesn't have any influence on each other.

- Social (geodesic) distance (denoted as $\mathbb{D}$): This distance can be easily computed by measuring the length of the short path between two users. The larger social distance between them indicates the less influence.
- Similarity between social context (denoted as $\mathbb{S}$): The similarity between social context can be measured by taking into account the matching patterns between the SNOs. It means the influence between two users from different social networks will be higher if their SNOs are more similar. Extending David and Euzenat [17], it is given by

$$\mathbb{S}(u_i, u_j) = \frac{CORR(\mathscr{O}_{SN_i}, \mathscr{O}_{SN_j})}{\max(|\mathscr{O}_{SN_i}|, |\mathscr{O}_{SN_j}|)} \tag{10.5}$$

where $\mathscr{O}_{SN_i}$ and $\mathscr{O}_{SN_j}$ are the SNOs by the two users, respectively. Also, $|\mathscr{O}|$ means a set of the ontology (i.e., the total number of only concepts and properties). The range of this similarity is $[0, 1]$.

Then, these two heuristic factors are applied to fuse social contexts with those from the other social networks.

Definition 10.5 (Fused context). A fused context of a user u is given in three different ways

$$FCTX_u = \sum_{u' \in SN_u^{\tilde{e}}} \mathbb{F}(CTX_{u'}, CTX_{u \to u'}^{\mathscr{S}}) \tag{10.6}$$

$$= \sum_{u' \in SN_u^{\tilde{e}}} \mathbb{F}\left(CTX_{u'}, \frac{CTX_{u \to u'}^{\mathscr{S}}}{\mathbb{D}(u, u')}\right) \tag{10.7}$$

$$= \sum_{u' \in SN_u^{\tilde{e}}} \mathbb{F}(CTX_{u'}, CTX_{u \to u'}^{\mathscr{S}} \times \mathbb{S}(u, u')) \tag{10.8}$$

in which two heuristics $\mathbb{D}$ and $\mathbb{S}$ have been applied to the fusion function $\mathbb{F}$. Here, the fusion function $\mathbb{F}$ is simply regarded as an union operation, since the score is equivalent.

As mentioned in [18, 19, 8], there can be a number of fusion functions. Thus, system administrators (or decision makers) can design more diverse fusion functions, depending on the business policies and strategies. A simple rule-based fusion method is discussed in Sect. 10.5 with a case study.

10.4   Mobile Services by Social Contexts: A Case Study

Once the fused context has been obtained by mobilizing social networks, one has to discuss how to exploit the context information to provide efficient services. In this section, we focus on applying the fused contexts to mobile recommendation service with a case study.

Assume that in an academic conference several sessions are in parallel. Various types of mobile services can be useful to the conference participants. We note the following potential scenarios.

Scenario 1 (Personalized scheduling). *Once the participants have made their registrations, they receive the full information about the conference. However, because such information is not personalized, they have to take a look at the entire conference program and make a decision whether each of presentations (or sessions) is relevant to their own research topics. The fused context of each participant can be comprised by ones previous and recent publications and research projects. Moreover, ones scholar (or coauthor) networks can be applied to enhance the fusion process. Then, the fused context can be exploited to be compared with a set of keywords from the research papers.*

Scenario 2 (Location-based service). *On the other hand, the conference provides opportunities for social interaction to exist between scientists. However, it is difficult to meet a specific person in sessions and social events. The fused context can include current location information through RFID tags or smart cards as well as coauthor networks, similar to the previous scenario. The participants can be efficiently informed of the location of the person who they wish to see.*

Scenario 3 (Social network management). *Even though participants have active social interactions with others in the conference, it is difficult to remember all of them without exchanging business cards. From the fused contexts , one can measure adjacency patterns among participants by using RFID tags or smart cards. The patterns are stored and merged into their own online social network systems, so that they can easily manage their social networks.*

10.5   Experimentation

To evaluate the proposed system, smart conference system has been implemented with RFID tags for all participants and RFID readers in all conference rooms. The system have been used in two Korean domestic conferences in 2009. The numbers of these conferences were 104 and 85, respectively. During registration, all the participants were asked to input their affiliation, OpenID[5], email address, and all accounts of major social networking systems. Social network datasets have collected from online social networking systems as well as offline social network activities.

- Online social networks

    1. (SN_1) FaceBook (http://www.facebook.com/)
    2. (SN_2) Twitter (http://twitter.com/)
    3. (SN_3) Cyworld [6] (http://cyworld.com/)
    4. (SN_4) DBLP (http://www.informatik.uni-trier.de/ ley/db/)

---

[5] http://openid.net/

[6] CyWorld is one of the most popular social network systems in Korea.

5. (SN_5) CiteSeer (http://citeseer.ist.psu.edu/)
6. (SN_6) Korean DBPIA (https://www.dbpia.com/)

- Offline social networks

1. (SN_7) Affiliation
2. (SN_8) Address

Though these user input, we constructed the ego-centric social network of each participant. Especially, DBLP , CiteSeer and DBPIA ontologies have been automatically matched with each other by OLA[7]. In practice, with OpenID and email address information, the redundant participants who have multiple accounts were detected .

Table 10.4 Performance of mobile recommendation at the first conference (%)

Parallel sessions	with Eq. 10.6			with Eq. 10.8		
	Session 1	Session 2	Session 3	Session 1	Session 2	Session 3
Morning	82.44	60.54	62.01	93.2	70.75	70.25
Afternoon 1	56.17	46.26	68.66	83.5	79.8	69.4
Afternoon 2	65.95	37.67	48.79	86.35	72.5	85.0

Table 10.5 Performance of mobile recommendation at the second conference with Eq. 10.8 (%)

Parallel sessions	Session 1	Session 2	Session 3	Session 4
Morning	68.95	85.3	91.2	85.5
Afternoon 1	84.24	77.95	79.33	89.4
Afternoon 2	84.3	89.25	84.5	85.25

During the conferences, the participants received the text messages occasionally via their cell phones occasionally. We focused on providing the following three mobile services to the participants and also evaluated them to justify whether the services work by measuring user satisfaction;

Recommendation of presentation schedule.   We have done a simple string matching between all papers (i.e., only keywords) in the conference proceedings and the fused contexts (i.e., $CTX^{\mathscr{I}}$) from SN_4, SN_5, and SN_6 with Eq. 10.6 and Eq. 10.8. For evaluation, we have computed the ratio of participants who have followed the mobile recommendation. We were easily able to find out which sessions the participants have been attended by referring to the RFID log database. The two one-day conferences were organized as 3 and 4 parallel sessions, as shown in Table 10.4 and Table 10.5.

---

[7] OWL-Lite Alignment [20]. http://ola.gforge.inria.fr/

Recommendation of potential users. Because we want them to interact as much as possible, we have sent a location of a particular participant to potential users during coffee breaks and banquet. The contexts from SN_1, SN_2, SN_3, SN_4, SN_7, and SN_8 have been fused by three different methods Eq. 10.6, Eq. 10.7 and Eq. 10.8. For evaluation, we have asked the participants to reply via SMS if they confront potential colleagues who have common research topics. The ratio of the replies to the recommendation has been measured at the conferences, as shown in Table 10.6. Furthermore, the time difference between the timestamps has been computed.

Table 10.6 Ratio of replies about recommending potential colleagues

	Eq. 10.6	Eq. 10.7	Eq. 10.8
Conference1	59.3% (= 51/86)	80.8% (= 42/52)	85.1% (= 40/47)
Conference2	60.3% (= 41/68)	78.3% (= 18/23)	85.3% (= 29/34)

10.6 Discussion

Two mobile services were conducted by us in academic conferences, and the experimental results collated. As the first service for recommending the personalized schedule, two participants groups were compared by using two different context fusion methods. Thus, we found out Eq. 10.8-based recommendation outperforms Eq. 10.6-based one by 40.3%. Overall, the average performances of recommendation services at those conferences have shown 78.97% and 83.76%, respectively. Through this, we have realized that social contexts extracted from the bibliographic social networks (i.e., DBLP , DBPIA , and CiteSeer) have been useful to compare contexts about research topics.

Secondly, in terms of recommending potential social colleagues, similarity between the social contexts (i.e., $\mathbb{S}$ in Eq. 10.8) has been the most useful information for context fusion, rather than others. Also, this potential social links have enriched online ego-centric social networks to foster efficient social collaborations between unknown people.

10.7 Summary

In this chapter, with two case studies, we have described a social network-based mobile recommendation services. This services are basically generated by taking into account meaningful dependencies between the contexts of two (or more) users. Given heterogeneous contexts extracted from multiple social networks, we have to take into account a context fusion process which can find out the hidden relationships between the contexts. From the experiments of both mobile services, we have realized that the SNO-based matching has been working well to discover meaning social relationships between the mobilized online social networks.

There have been several research limitation. In the first case study, even though we have relatively enough dataset to discover a social network, we were not about to evaluate the results because of the privacy issue on the subscribers. Moreover, in semantic web communities, there have several efforts to standardization metadata to represent a digital identity and an online social network (e.g., FOAF, SIOC, and so on). As future work of this study, we are planning to conduct the following issues;

- detection of dynamic change, and
- uncertainty reasoning for social context fusion

Acknowledgements. This work was supported by the National Research Foundation (NRF) grant funded by the Korean government (MEST) (2009-0066751).

References

1. Sekkas, O., Anagnostopoulos, C.B., Hadjiefthymiades, S.: Context fusion through imprecise reasoning. In: Proceedings of the 2007 IEEE International Conference on Pervasive Services (ICPS), pp. 88–91. IEEE Computer Society, Los Alamitos (2007)
2. Cabri, G., Leonardi, L., Mamei, M., Zambonelli, F.: Location-dependent services for mobile users. IEEE Transactions on Systems, Man, and Cybernetics - Part A: Systems and Humans 33(6), 667–681 (2003)
3. Korpipaa, P., Mantyjarvi, J., Kela, J., Keranen, H., Malm, E.-J.: Managing context information in mobile devices. IEEE Pervasive Computing 2(3), 42–51 (2003)
4. Herrera, F., Martínez, L.: A model based on linguistic 2-tuples for dealing with multi-granular hierarchical linguistic contexts in multi-expert decision-making. IEEE Transactions on Systems, Man, and Cybernetics - Part B: Cybernetics 31(2), 227–234 (2001)
5. Melgani, F., Serpico, S.B., Vernazza, G.: Fusion of multitemporal contextual information by neural networks for multisensor image classification. In: Proceedings of the 2001 IEEE International Geoscience and Remote Sensing Symposium (IGARSS 2001), pp. 2952–2954. IEEE Computer Society, Los Alamitos (2001)
6. Pant, G., Srinivasan, P.: Link contexts in classifier-guided topical crawlers. IEEE Transactions on Knowledge and Data Engineering 18(1), 107–122 (2006)
7. Jung, J.J.: Ontological framework based on contextual mediation for collaborative information retrieval. Information Retrieval 10(1), 85–109 (2007)
8. Jung, J.J., Lee, H., Choi, K.S.: Contextualized recommendation based on reality mining from mobile subscribers. Cybernetics and Systems 40(2), 160–175 (2009)
9. Irving, R.H., Conrath, D.W.: The social context of multiperson, multiattribute decision-making. IEEE Transactions on Systems, Man and Cybernetics 18(3), 348–357 (1988)
10. Cross, R., Rice, R.E., Parker, A.: Information seeking in social context: Structural influences and receipt of information benefits. IEEE Transactions on Systems, Man, and Cybernetics - Part C: Applications and Reviews 31(4), 438–448 (2001)
11. Eagle, N., Pentland, A.: Reality mining: sensing complex social systems. Personal and Ubiquitous Computing 10(4), 255–268 (2006)
12. Erétéo, G., Buffa, M., Gandon, F., Corby, O.: Analysis of a real online social network using semantic web frameworks. In: Bernstein, A., et al. (eds.) ISWC 2009. LNCS, vol. 5823, pp. 180–195. Springer, Heidelberg (2009)

13. Alani, H., Szomszor, M., Cattuto, C., van den Broeck, W., Correndo, G., Barrat, A.: Live social semantics. In: Bernstein, A., Karger, D.R., Heath, T., Feigenbaum, L., Maynard, D., Motta, E., Thirunarayan, K. (eds.) ISWC 2009. LNCS, vol. 5823, pp. 698–714. Springer, Heidelberg (2009)
14. Euzenat, J., Shvaiko, P.: Ontology matching. Springer, Heidelberg (2007)
15. Dey, A.K.: Understanding and using context. Personal and Ubiquitous Computing 5(1), 4–7 (2001)
16. Jung, J.J.: Contextualized mobile recommendation service based on interactive social network discovered from mobile users. Expert Systems with Applications 36, 11950–11956 (2009)
17. David, J., Euzenat, J.: Comparison between ontology distances (preliminary results). In: Sheth, A., et al. (eds.) ISWC 2008. LNCS, vol. 5318, pp. 245–260. Springer, Heidelberg (2008)
18. Hofreiter, B., Huemer, C., Winiwarter, W.: Business collaboration models and their business context-dependent web choreography in BPSS. International Journal of Web Information Systems 1(1), 33–42 (2005)
19. Bessai, K., Claudepierre, B., Saidani, O., Nurcan, S.: Context-aware business process evaluation and redesign. In: Nurcan, S., Schmidt, R., Soffer, P. (eds.) Proceedings of the 9th Workshop on Business Process Modeling, Development and Support in conjunction with the CAiSE 2008 conference, Montpellier, France, June 16-17 (2008)
20. Euzenat, J.: An API for ontology alignment. In: McIlraith, S.A., et al. (eds.) ISWC 2004. LNCS, vol. 3298, pp. 698–712. Springer, Heidelberg (2004)

Index

Retrieval accuracy   102
Reward Function for Session   39
RFID   260, 261

Sampling Error   64
Search Engine Optimization   127
Semantic
   Applications, 193
   Clustering, 154
   Representation Model, 178, 186
      Environmental information, 178
      Execution type, 178
      Input/output, 179
      Level of familiarity, 180
      Location, 178
      Low level context, 178
      post-conditions, 179
      preconditions, 179
      Space constraints, 179
      Space level of publicness, 179
      Space potential, 179
      Time, 178
      User interaction type, 178
      User preferences, 179
      User type, 178
   SOA, 213, 214
   Technology, 210, 212, 214
   Web, 193, 219
SEO   127
Service Oriented Computing   168
   Service discovery, 176, 181
   Service query manager, 177
   Service scheduler, 177
Session
   Outliers, 41
   Representation, 150
Sessionization   148
Singular Variable Decomposition
      (SVD)   58
SMS   262
SOA   214
Social
   Affinity, 252
   Context, 252, 258
   Network, 252, 253
      Mobilization of, 254
      Ontology (SNO), 254, 257, 258
SOFM   34, 62, 154
Spam   127

SPARQL, query language   183
Spontaneous task-composition   187
Squared Error
   Mean, 66
   Relative, 66
   Root Mean, 66
Statistical Stemming   30
Stemming   30
Strategic Adaptation   159
Strongly Connected Component   115
Structural Clustering   128
Supervised Learning   59
Support Vector Machines (SVM)   59
SVM
   Outliers, 27

Tactical Adaptation   159
Task-oriented
   Computing, 168, 186
      Abstract Services, 169, 176, 181
      Actions, 169, 175, 186
      Service Composition Pattern
         (SCP), 169, 175, 181
      Tasks, 168, 175, 178, 186
      Users' Goals, 168, 175, 186
   Service Framework, 168, 175, 176
      Application Template, 175, 186
      Composition-pattern Layer, 169,
         175, 187
      Description Manager, 176
      Interpreter, 176
      Property Matchmaker, 176, 182
      Semantic Measurer, 176
      Service Composition-pattern
         Ontology, 176
      Service Layer, 169, 187
Teleportation   124
Temporal Evolving Graph   23
Text Weighting Schema   30
Time-use Studies   178, 183, 187
TKC   128
Tracking Application   36
Transposed Matrix   114
TrustRank   128
Twitter   260

Ubiquitous Computing   168
   Environment, 168, 170
   Frameworks, 173
      ABC, 174